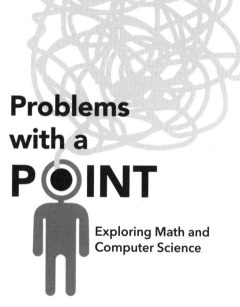

Problems with a P**O**INT

Exploring Math and Computer Science

William Gasarch | Clyde Kruskal

University of Maryland, USA

Problems with a

P●INT

Exploring Math and Computer Science

World Scientific

NEW JERSEY · LONDON · SINGAPORE · BEIJING · SHANGHAI · HONG KONG · TAIPEI · CHENNAI

Published by

World Scientific Publishing Co. Pte. Ltd.

5 Toh Tuck Link, Singapore 596224

USA office: 27 Warren Street, Suite 401-402, Hackensack, NJ 07601

UK office: 57 Shelton Street, Covent Garden, London WC2H 9HE

British Library Cataloguing-in-Publication Data
A catalogue record for this book is available from the British Library.

PROBLEMS WITH A POINT
Exploring Math and Computer Science

ISBN 978-981-3279-72-8
ISBN 978-981-3279-97-1 (pbk)

For any available supplementary material, please visit
https://www.worldscientific.com/worldscibooks/10.1142/11261#t=suppl

Desk Editor: Amanda Yun

Printed in Singapore

Preface

In 2003 Lance Fortnow started a blog on complexity theory, aptly titled *complexityblog*. I was a frequent guest blogger and in 2007 became a co-blogger. While the topic was ostensibly complexity theory, we both also blogged about math, CS, education, math education, CS education, academic politics, real politics, and really anything we felt like.

One of my favorite types of humor is observational, for example, Jerry Seinfeld has said:

> *Did you ever notice, when you are sitting at a red light, that when the person in front of you pulls up a couple of inches, you are compelled to move up too? Do we really think we making progress towards our destination? "Whew, I thought we would be late, but now that I am nine inches closer, I can stop for coffee and a danish."*
>
> *[audience laughs]*

One of my favorite types of blog posts is to make an observation, Jerry-Seinfeld-like, and then present some math to illustrate that point. For example, in Chapter 24, we expand on the following observation:

> *Did you ever notice how sometimes you know a statement is false since if it was true you would have heard about it? [audience laughs]*

This gives us the alliterative title: *Problems with a Point.*

Format of the Chapters; Format of the Book
This book is a collection of essays inspired by some of my blog entries. In some cases a blog entry was one page and the essay is over ten pages. Additionally these essays are polished and include more details and background material.

My original vision for the book was that *every* chapter would begin with a point about math and then present the math to illustrate that point. I quickly found, as Ralph Waldo Emerson said:

> *A foolish consistency is the hobgoblin of small minds*

While some chapters follow that format, some do not.

The chapters are partitioned into three parts which are arranged roughly in order of mathematical sophistication. A more precise measure of difficulty is provided by having a note on *Prior Knowledge Needed* at the beginning of every chapter.

(1) *Stories with a Point:* Each chapter in this part illustrates its point by telling stories related to mathematics. There are no proofs.
(2) *Problems with a Point:* Each chapter in this part illustrates its point by challenging the reader with a problem and then presenting the solution.
(3) *Theorem with a Point:* Each chapter in this part illustrates its point by presenting theorems and their proofs.

The Elephant in the Room
So why should you buy this book if the blog posts are available for free?

(1) Trying to find which entries are worth reading would be hard.

(2) The description above portrays my blogs as well thought out. In reality the notion that I have a point to make and use math to make that point is revisionist history. The chapters in this book are polished and expanded versions of blog entries.

(3) There is something about a book that makes you want to read it. Having words on a screen just isn't the same. I used to think this was my inner Luddite talking, but it turns out that most younger people that I've spoken to agree, especially about math-on-the-screen.

(4) Lance hopes that I'll fix the spelling, grammar, math, CS, etc. Lance even recommended I get a co-author to help me refine, polish, etc. So I did: Clyde Kruskal. Even so, the book is written in my voice.

What background does the reader need?

A few chapters need no math at all! Most chapters need the mathematical maturity of a bright high school student. Some chapters need more. Every chapter has, after the title, a note about what prior mathematics is needed.

Acknowledgments

I thank Lance Fortnow for giving me a chance to guest-blog and later co-blog. He was the blogger in the area of computer science theory, and taught me many lessons about blogging. The most important one is to not play-to-the-crowd, i.e. don't let comments discourage or guide what you do. As Rick Nelson taught us in his song, *Garden Party:*

You can't please everyone, so you got to please yourself.

I thank Rochelle Kronzek who encouraged me to write this book and helped to shape it. I thank the internal reviewers of the book who gave valuable advice on some of the chapters and who also advised me to get a co-author.

I thank my Darling for being a good sounding board and putting up with me talking about the book. She has a Masters in Software Engineering. Therefore she knows *enough* math to know what I am talking about but *not enough math* to have drunk the Kool-aid. She often calls me out on the absurdity of some of what mathematicians and CS theorists do. You *do not* want to get into an argument with her about the Banach-Tarski Paradox. She claims it means math is broken. She might be right.

Clyde thanks his significant other Anne Laurie Wong for editing some chapters and her patience.

Four books inspired me.

(1) **Quantum Computing Since Democritus,** by Scott
    Aaronson. While ostensibly about quantum computing,
    there are chapters on *The Anthropic Principle, Time Travel,*
    and other topics that are not encompassed by the title. This
    taught me to either not be bound by my title, or to get a ti-
    tle that covers just about everything (I leave it to the reader
    to determine which I did). In addition, Scott's fine writing
    is an inspiration.

(2) **People, Problems, and Proofs,** by Richard Lipton and
    Ken Regan. The book is based on their computer science
    blog, that is, frankly, more about math than computer sci-
    ence. Their breadth of knowledge and ability to write clearly
    encouraged me to try to do the same. Their book is a proof-
    of-concept for my book.

(3) **Professor Stewart's Cabinet of Mathematical Cu-
    riosities,** by (who else?) Ian Stewart. This book is a proof
    by example of Emerson's quote in the preface. The chapters
    have all kinds of different formats and lengths, yet the book
    works. Not *despite* the lack of uniformity, but *because* of it.

(4) **Structure and Randomness: Pages from Year One
    of a Mathematical Blog,** by Terence Tao. This is a blog-
    inspired book based on his math blog. *One sentence* in it
    struck me, which I paraphrase:

    > *My proof of Hilbert's Nullstellensatz is probably
    > not new but I thought I would share it here any-
    > way.*

    This excerpt inspires me to make sure every chapter is worth
    sharing with my readers.

Contents

PART 1

Stories with a Point

Chapter 1

When Terms from Math are Used by Civilians...

Prior Knowledge Needed: None.

1.1 Point

Do civilians use math words correctly? Am I using the word *civilian* correctly? The word *civilian* used to mean *non-military*; however, I have seen it to mean *non-X* for some values of X.

Which math words are entering the mainstream? When a civilian uses a math word correctly does it educate the public? When a civilian uses a math word incorrectly does it educate the public since a listener might look it up?

We look at several math words that have been used by civilians and try to answer some of these questions.

1.2 $n + 1$

There is a general interest magazine called $n + 1$ [Anonymous (2004–)]. I emailed the editors to ask what the name means and I got the following enlightening response:

> *Well, as a non-math person and one of the founding editors of $n + 1$, I can tell you that we still don't know very much about math, but we did have some vague high school memories of set theory and algebra and knew that*

3

n + 1 could mean, if it were a set, an infinite series or open-ended expansion, or just that for any quantity (n), there's often more than meets the eye, or is commonly thought or known (+1). That was the sense that Chad Harbach, another founding editor, had in mind when he first thought of the title as a placeholder, a math metaphor for human potential, back when he was a Harvard undergrad. Over time, the title also seemed to work in response to the End of History crowd, all those people who told us that no new ideas were really possible in the humanities, no new writing was possible, that it was foolish to start a magazine on politics, literature, and culture in these times. So we took on n + 1 as a rallying cry, of sorts. Someone might also hear it as end+1 after the end, a new beginning, that sort of thing. We did have a math PhD friend who suggested that, if we really wanted to designate an infinite universe of possibilities, we should have called it ω + 1, but that seemed too much for us non-math types. As far as I know, ω + 1 is still available as a title.

Thanks for writing in to ask and best of luck with the blog and other endeavors,

I wish them well too!

Is this a proper usage of the phrase $n + 1$? Well... its not an improper use of it. Will the use of $n + 1$ educate anyone? Probably not. Here's hoping it does not scare away math-phobic readers.

1.3 Quantum

The *Quantum* in *Quantum Mechanics* means *Discrete Quantity*. It does not necessarily mean *small quantity*. We give some examples of its use.

Rick and Morty is a cartoon TV show where Rick is an excellent (though alcoholic) scientist and Morty is his 14-year old grandson. In the episode *The Rick Must be Crazy* their car broke down and they had the following conversation

Morty: *W-Whats wrong Rick? Is it the Quantum carburetor or something?*

Rick: *Quantum carburetor? You can't just add a Sci-fi word to a car word and hope it means something.*

Alas – if only other writers would take Rick's advice.

Quantum Leap is the name of a TV show where Sam Beckett (a fictional character, not the playwright) bounced around time inhabiting people's bodies and setting things right. The leaping from one body to another is a *Quantum Leap*. Josh Jones, a fan of the show (I would think), posted an interactive map [Jones (year unknown)] of all the leaps Sam Beckett experienced in the TV series.

Is the term *Quantum Leap* being used correctly? I originally thought *no* since I thought "Quantum" meant *large*. Having looked at the web for non-tv uses of the term I have changed my mind; it is a correct use of the term. Here is what I found on the web:

(1) A *Quantum Leap* is *the discontinuous change of the state of an electron in an atom of molecule from one energy level to another.*

(2) A *Quantum Leap* is an abrupt extreme change (this usage best fits the TV show).

(3) *Quantum Leap* is the name of a restaurant which serves *Meatless fare with Asian and Mexican accents plus sweets like soy milkshakes and vegan pancakes.*

(4) *Quantum Leap* is the name of a healthcare collaborative.

(5) *Quantum Leap* is used a lot for ads promising an improvement in the client's life (if they mean a *large* improvement, then this is incorrect use. If they mean an *abrupt* change then this is a correct use. I give them the benefit of the doubt).

Quantum of Solace is the name of a James Bond short story by Ian Fleming, and of a James Bond movie. The short story and the movie only overlap in that they both have James Bond in them.

Quantum has a double meaning in the movie. They use Quantum to mean both (1) an amount of solace, and (2) the name of the criminal organization that James Bond is fighting. Kudos to the writers for using the term correctly, although in an absolutely awful movie.

Quantum Theater is ... I give excerpts from their website [Quantum Theatre] rather than try to paraphrase:

> *Quantum artists mine all kinds of non-traditional spaces for the sensory possibilities they offer when combined with creative design.*

> *We find it meaningful to place the audience and performers together, the moving parts inside the work.*

> *The shows run the gamut from those you thought you knew but now experience like never before, to shows that didn't exist until their elements mixed in our laboratory.*

If you know what any of this means then let me know.

1.4   Prisoner's Dilemma

1.4.1   *An Article in The New Republic*

In the June 17, 2009 issue of *The New Republic* there is an article by Gabriel Sherman [Sherman (2009)] on Bob Woodward writing a book about the Obama White House. The article is titled *Plan of Attack*, though if could have been titled *Prisoner's Dilemma* (we will later see why it was not). Here is an excerpt:

> *He (Bob Woodward) flashes a glimpse of what he knows, shaded in a largely negative light, with the hint of more to come, setting up a series of prisoner's dilemmas in which each prospective source faces a choice: Do you cooperate and elaborate in return for (you hope) learning more and earning a better portrayal — for your boss and yourself? Or do you call his bluff by walking away in the hope that your reticence will make the final product less authoritative and therefore less damaging? If no one talks, there is no book. But if someone talks, then everyone always talks.*

Kudos to the article for using the term *Prisoner's Dilemma* correctly! Curiously they never defined the term; however, I suspect the reader can get it from context or (in the modern electronic age) easily look it up. In this way the article may educate the public in ways beyond its content.

The article right after *Plan of Attack* was titled *Prisoner's Dilemma*. It was about what to do with the detainees in Gitmo. That article only used the term *Prisoner's Dilemma* in the title. Since what to do about the prisoners in Gitmo was a dilemma (and still is) the title is correct usage. However, its use as a title here meant it could not be used as a title for the article about Woodward. Too bad. *Plan of Attack* is a terrible title.

1.5   Turing Test

Roger Ebert's review of the movie, *X-men Origins: Wolverine*, is titled

A monosyllabic superhero who wouldn't pass the Turing Test.

In the review he never mentions the Turing Test or what it means. Does he expect his readers to know? Do they? Will readers of the Ebert's column go to Wikipedia to find out? Might this be a way to educate people?

Having seen the movie, I can say with confidence that Ebert used the term correctly.

1.6   Venn Diagram

1.6.1   *Tiger Woods*

In 2010 Tiger Woods was involved in a sex scandal. Lisa Taddeo [Taddeo (2010)] wrote about the world Tiger Woods was involved in. The following passage compared options for men who have affairs, whether with a civilian or prostitute:

> *Both methods of slaking the hunger have their pros and cons. Men like to hunt, and there is no need to hunt a prostitute. Men like to cheat without strings, and you can't stop a civilian from falling in love but (Tiger) Woods found a way to enjoy the best of both worlds in one type of women. A* Venn diagram *of sexual satisfaction. Most of his women lived in a nebulous in-between world.*

Will this enlighten the masses as to what a Venn Diagram is? If they look it up then yes; however, the actual statement is wrong. What they really mean is *an intersection of sexual satisfaction*, or as it's commonly known, an *intersextion*.

1.6.2   *Speaker of the House*

When John Boehner stepped down as speaker of the house, in 2015, there was speculation over who would replace him. Matt Lewis [Lewis (2015)] wrote an article for *The Daily Beast*, which we excerpt:

> *Option 3: An acceptable and respected conservative like Jeb Hensarling or Tom Price emerges as speaker. Why these two? First, Paul Ryan doesn't seem to want the gig, so that leaves us with only a few options for some-one who fits the* Venn diagram *of being enough of an outsider, well liked, and sufficiently conservative.*

They really mean the *intersection* of an outsider, well-liked, and sufficiently conservative. I would not say their usage is wrong. Even though Paul Ryan became speaker, I would not say their statement *Paul Ryan doesn't seem to want the gig* was wrong.

1.7   Zeno's Paradox

In 2010, plans were made for the last Harry Potter book to be made into two movies. David Edelstein [Edelstein (2010)] wrote about the decision:

> *When Warner brothers announced that the seventh and final book of J.K. Rowling's Harry Potter series,* Harry Potter and the Deathly Hallows, *would be two movies, it occurred to me that the company had been insufficiently ambitious. If, as reported, Warner executives are scared of running short of tentpoles (i.e., the so-called*

franchises that prop up a studio), they should at the very least divide the next half in half. Following Zeno's paradox, they could even turn Deathly Hallows into an infinite number of sequels with Beckett-like arcs of non-action: Let's apparate. [They do not move.]

Is this the correct usage of Zeno's paradox? Beckett? Apparate? Will it inspire people to learn about Zeno's paradox? It inspired me to look up "apparate" in the *Urban Dictionary*:

Apparate: *In the Harry Potter world, this is a spell which allows the user to instantaneously teleport from one location to another.*

1.8   Have Any of These Phrases Entered the Mainstream?

First lets check how many Google hits each of the phrases gets. I round-off, since a statement like

"William Gasarch" gets 7782 hits

implies too much precision. This is partly because if someone Googled "William Gasarch" tomorrow they may well get a different numbers of hits.

In order of number of Google Hits (from highest to lowest):

(1) *Quantum* gets around 185,000,000 hits.
(2) *Venn diagram* gets around 7,000,000 hits.
(3) *Turing test* gets around 1,000,000 hits.
(4) *Prisoner's Dilemma* gets around 900,000 hits.
(5) *Zeno's Paradox* gets around 82,000 hits.

Quantum and *Venn Diagram* are already in the mainstream; however, they are often used incorrectly. *Turing Test* and *Prisoner's Dilemma* may enter the mainstream and may even be used correctly. I was surprised that *Zeno's Paradox* appeared in the popular press; I doubt it will ever enter the mainstream.

References

Anonymous (2004–). What we are, `https://nplusonemag.com/about`.

Edelstein, D. (2010). Harry Potter and the dragged out final act, *New York Magazine* `http://nymag.com/movies/reviews/69478/`.

Jones, J. (year unknown). Oh boy: The complete quantum leap journey, `https://specialrequest.media/Oh-Boy`.

Lewis, M. (2015). What's next after John Boehner, *The Daily beast* `https://www.thedailybeast.com/whats-next-after-john-boehner`.

Quantum Theatre (1990–). What we are, `http://www.quantumtheatre.com/about/what-we-are/`.

Sherman, G. (2009). Plan of attack, *The New Republic* `https://newrepublic.com/article/63664/plan-attack`.

Taddeo, L. (2010). Rachel Uchitel is not a madam, *New York Magazine* `nymag.com/news/features/65238/`.

Chapter 2

Sports Violate the Rules of Mathematics!

Prior Knowledge Needed: None.

2.1 Point

The title of this chapter is the point of this chapter.

2.2 Baseball Violates the Pythagorean Theorem: Home Plate

Wolfram's MathWorld and its sources claim the rules of baseball say that home plate is a 17.5 inch rectangle with a 12-12-17 right triangle on top of one of the long sides. This can't be quite right since there is no 12-12-17 right triangle. What do they really mean? Here are some options:

- If Major League Baseball (MLB) insists on using a right triangle with legs of length 12 and 12 then the hypotenuse is of length $\sqrt{12^2 + 12^2} = \sqrt{288} = 12\sqrt{2}$. This is approximately 16.97 which is very close to 17.
- If MLB insists on using an isosceles right triangle with hypotenuse 17 then the legs must satisfy the equation $2L^2 = 17^2$, so $L = \frac{17}{\sqrt{2}}$. This is approximately 12.02 which is very close to 12.

- If MLB insists on using an isosceles triangle with base 17 and legs 12 then the angle in question is of size is $2\sin_{-1}\frac{8.5}{12}$ degrees. This is approximately 90.2 which is very close to 90.

Which is the best to use? That is a matter of opinion. Which one is used? That is a matter for speculation. Or I could just buy a home plate and measure it, though I doubt I could tell which is true.

In preparing this chapter I wondered if this is really the definition. *Wolfram's MathWorld* and its sources claim that MLB insists on a 12-12-17 right triangle, and note that this is impossible. I then looked in the MLB rule book.

> *Home base shall be marked by a five-sided slab of whitened rubber. It shall be a 17-inch square with two of the corners removed so that one edge is 17 inches long, two adjacent sides are 8.5 inches and the remaining two sides are 12 inches and set at an angle to make a point. It shall be set in the ground with the point at the intersection of the lines extending from home base to first base and to third base; with the 17-inch edge facing the pitcher's plate, and the two 12-inch edges coinciding with the first and third base lines. The top edges of home base shall be beveled and the base shall be fixed in the ground level with the ground surface. (See drawing D in Appendix 2.)*

They never quite say that it's a right angle. But they do say that the lines extend to first base and third base which *implies* a right angle. The picture in Appendix 2 has home plate snugly in a corner of a square which also *implies* a right angle. According to the rules they seem to use a 12-12-17 triangle and it's okay that there is no right angle.

Therefore home plate violates mathematics!

2.3 Baseball Violates The Laws of Statistics: Batting Averages

A player's *batting average* is the percent of at-bats for which he or she gets a hit. Note that the following do not count as at-bats:

(1) The batter receives a walk.
(2) The batter is hit by a pitch.
(3) The batter hits a sacrifice fly or sacrifice bunt.
(4) The batter is given first base because an opposing player, usually the catcher, interfered. Usually? That's what Wikipedia says; however, I cannot imagine any other player interfering.
(5) The player is replaced at the plate by a pinch hitter. There is one exception. The pinch hitter will inherit the count from the first batter. If the count has 2 strikes, and the pinch hitter strikes out, then *the first batter* is charged with the at-bat and the strikeout. Is that fair? I don't know. I have never seen a player replaced during his at-bat; however, if baseball statistics get more refined we may see players that specialize in hitting with (say) a 1-2 count.
(6) If the inning ends before the batter gets to finish his at-bat (e.g., a player tries to steal a base and is thrown out for the third out of the inning). In this case the batter comes back at the beginning of the next inning with a count of 0-0.

We will leave out the decimal point and hence write 237 for a batting average rather than .237.

Mario Mendoza was a shortstop who was an excellent fielder but a terrible hitter. So much so that *The Mendoza Line* is an expression to mean the line below which no matter how good a fielder you are, you have no place on a major league roster. The term is so common that (1) it gets 63,000 hits on Google,

(2) there is a movie with that title, (3) there is a rock group with that name.

Let's say that the Mendoza line is 200. Then it is not surprising that there are *more* players batting 200 then 199. And 201 then 200. But once you are above (say) 260, which is a respectable batting average, the higher you go the *fewer* people have that average. More concretely, in 2009 there were 432 players who had at least 100 at-bats and the batting average was approximately normally distributed with mean 261 and standard deviation 34. This kind of normal distribution for batting average is typical of baseball history.

Let $N(a)$ be the number of players with a batting average of a over all of baseball history. If $N(a)$ was normally distributed with mean roughly 260 then one would expect the following:

$$N(296) \geq N(297) \geq N(298) \geq N(299) \geq N(300) \geq \cdots$$

Lets look at the data (as of 2009)

(1) $N(296) = 123$
(2) $N(297) = 139$
(3) $N(298) = 128$
(4) $N(299) = 107$
(5) $N(300) = 195$
(6) $N(301) = 157$
(7) $N(302) = 156$
(8) $N(303) = 123$
(9) $N(304) = 147$

Oh! There are far more players batting 300 than 299. Or 298. Or 297. Or 296. This would seem to violate statistical theory! And common sense! And baseball! Or not! Also note that the dropoff from 300 to 301 is larger than the dropoffs from 301 to 302, 302 to 303, or 303 to 304. If you are a baseball fan you can probably guess the explanation: Batting 300 has become a standard that players try to achieve. Assume YOU are a major league ballplayer. If you are batting 300 and it is the last week of the season you may become very selective on what balls you hit. Hence you may walk more but get fewer outs and fewer hits. Or you may ask to sit out a game to preserve your batting average. Similarly, if you are hitting 296–299 you will do what you can to get a hit, perhaps settling for singles (low-risk, low-reward) rather than trying to get Home Runs (high-risk, high reward). Note that you are doing this for personal glory; it might hurt the team. Shame on YOU!

Bill James, the father of modern baseball statistics, calls this *The Targeting Phenomena*. He has observed that it happens with batting averages (targets: all $x \equiv 0 \pmod{10}$, but 300 especially), number-of-hits (target: all $x \equiv 0 \pmod{10}$, but 200 especially), runs batted in (target: all $x \equiv 0 \pmod{10}$, but 100 especially), strikeouts for a pitcher (200, 300), and wins for a pitcher (20).

Bill James has also shown, using statistics, that the Targetting Phenomena didn't really start until about 1940. He speculates that the Hall of Fame opening up in 1939 may have partially caused it.

2.4   Target: 3000 Hits in a Career

This also happens for lifetime records with (1) players hanging on to get 2000 or 3000 hits, and (2) pitchers hanging on to get 200 or 300 wins. We consider these players in this section.

The rule is

> *Players strive to get at least 3000 hits.*

Lets look at

> *the exceptions that TEST the rule*

which is how the expression *the exception that PROVES the rule* should be stated and understood. If you have a rule and there are some exceptions, but each one has a good explanation, then the exceptions have tested the rule.

Lets look at the players that had just under 3000 hits in order of how many hits they had. Recall that Bill James showed players did not care about statistics that much until about 1940, which accounts for some of the exceptions below.

(1) Stan Ross retired in 1995 after making his 3000th hit (or so he thought). He opened a chain of restaurants called *Mr. 3000*. But it turns out there was a clerical error and he only had 2997 hits. So he joined the 2004 Milwaukee Brewers just to get the three hits he needed. Did he succeed? See the movie *Mr. 3000* for the conclusion of this fictional story. Or look it up on Wikipedia.

(2) Sam Rice, 2987 hits, retired in 1934, before 1940.

(3) Sam Crawford, 2961 hits, retired in 1917, before 1940.

(4) Frank Robinson, 2943 hits, retired in 1976. He needed 57 more hits to get to 3000. His 1976 season was mediocre: 36 games, 15 hits, 224 batting average, 3 home runs, 10 RBI's, 10 walks. In addition, in his last season, he was a player-manager, so he may have put the needs of the team ahead of his own personal glory. It is unlikely he could have gotten 3000 hits.

(5) Barry Bonds, 2935 hits, retired in 2007. He also had a batting average of 298, 1996 Runs Batted in, and 762 Home Runs. His 2007 season was pretty good: 126 games, 94 hits, 276 batting average, 28 Home Runs, 66 RBI's, 132 walks (the number of walks lead the league). He certainly could have played another season and likely would get 65 hits (for a total of 3000), 4 Runs Batted (for a total of 2000). 38 Home Runs (to get to 800) is doubtful, but if he played two more seasons then it's plausible. His lifetime batting average may have gone down a bit. So why did he retire? No team wanted to sign him. He sued claiming collusion, but he lost. We can only speculate, however, that (1) he was expensive, (2) he was involved in a steroids scandal, and (3) he was considered difficult to work with.

(6) The next three players all retired before 1940: Willie Keeler (2932 hits, retired 1910), Jake Beckley (2930 hits, retired 1907), Rogers Hornsby (2930 hits, retired 1937).
(7) Al Simmons (2927 hits, retired 1944). Needed 73 hits. Without going into detail, his last two seasons indicate that he was not able to play major league baseball anymore.
(8) The next two players all retired before 1940: Zach Wheat (2884 hits, retired 1927), Frankie Frish (2880 hits, retired 1937).

Interestingly, Roberto Clemente had exactly 3000 hits. He didn't plan that: he died in a plane accident in December 1972.

2.5 Target: 300 Wins in a Career

I won't go into detail on this one. Suffice to say, the pitchers who were just below 300 wins either were before 1940 or really were past their prime. Some had arm troubles.

Exercise

(1) Assume we had six fingers on each hand. What would the targets be?
(2) Assume we had x fingers on each hand. What would the targets be?

2.6 The (Il)logic of Fandom

In 2013, the Sunday before the Superbowl, I spotted a curious passage in *The Washington Post* which I paraphrase:

> *Washington Redskins fans should root for the Baltimore Ravens in the Superbowl because the two teams share many fine qualities. Both have gritty defense, soft-spoken but shrewd head coaches, and underrated*

quarterbacks craving validation on the big stage. Nei-
ther team is flashy or has big stars — rather they are
both greater than the sum of their parts.

I interpret this as assuming sports fans pick who to root for
in a logical manner. Perhaps like this:

(1) Bill lists qualities he likes in a sports team.
(2) Bill looks at all of the teams and determines which one best
    meets his criteria (this may involve some mathematics).
(3) Bill will root for that team.

Does Bill do this? Does any fan do this? Do fans who are
mathematically inclined do this? The answers are no, no, and
no.

So what is the logic behind fandom? Why do people root for
certain teams, likely for a lifetime? Is it rational? We enumerate
some reasons people root for the teams they do.

(1) Root, root, root, for the home team, for if they don't win,
    it's a shame.
(2) Root for your childhood team.
(3) Root for a team that you can see on TV. I wonder if the
    Chicago Cubs have more fans since one of the early Cable
    Stations, WGN, is from Chicago and (I think) carries Cubs
    games.
(4) Undergraduates often root for their college team. Graduate
    students often root for themselves to get a PhD in $\leq X$
    years for some X.
(5) Fair weather fans root for their home team when the team
    is winning, but ignore the team at other times.
(6) The early NY Mets (early 1960's) were a terrible team but
    they had lots of fans. There was a loser-chic factor. The
    Chicago Cubs have also had that. This seems to be rare or
    even nonexistent these days.

(7) Individual players may have an effect. Tim Teabow got some (mostly positive) attention because of his devout Christianity. Ray Rice got some (mostly negative) attention because he beat up his fiancé (now wife) on a tape that was widely seen. Liking or disliking an individual player may make you like or dislike a team.

(8) A while back some teams delayed getting any black players on them so they would appeal to racist fans. This stopped once such teams just lost too much since they were not using that talent pool. Might someone root for or against a team based on how many players on it are black? White? Jewish? (Unlikely — there are more Jewish football owners than football players.) Christian? Muslim? Gay? Straight? Nonbinary? Some combination of these factors? Personally, I root for the team with the most left-handed Latino lesbians.

(9) Might someone like or hate a team based on the politics of the management? If some team took the profits and funded alternative energy for real would you root for them? Replace *alternative energy* with any of the following:

(a) Republicans
(b) Democrats
(c) "Science" to show that global warning is a hoax
(d) Ramsey Theory

I doubt a team would have a public opinion on an issue which may cause them to lose fans. And of course all teams want to avoid a repeat of the scandal when it was discovered that the Chicago Bulls owner funded research in Recursive Algebraic Topology.

(10) A friend of yours is on the team so you root for the team and your friend. It may be enough if someone on the team is from your school. Whenever I hear that some pro football player is from Harvard (where I went to graduate school) I take note. This is a rare event.

(11) Jerry Seinfeld once commented that we LOVE this player if he's on OUR team but HATE him if he is on another team. What has changed? His uniform. So we are rooting for clothing.

Why root, root, root for the home team?

From 1976–1981 and 1986–1989 the NY Mets had a player named Lee Mazzilli who was born in Brooklyn (NY). They played in Shea Stadium which is in Queens (NY). I saw a big poster advertising *Come see this Brooklyn boy play his heart out in Queens!* They were trying to brag about the fact that *one* of their players was actually from New York. Rather curious since they were also sending the message that *twenty four* of their players were not from New York. (He also played for the Texas Rangers (1982), the NY Yankees (1982), the Pittsburgh Pirates (1983–1986), and the Toronto Blue Jays (1989). He managed the Baltimore Orioles (2004–2005) and was an announcer for the NY Yankees (2000–2003, 2005). The NY Yankees could brag he was from NY, but the other teams could not.) So there is no real connection between the team *The NY Mets* and the place *NY*.

Lets look at college sports for whether it makes sense to root for a team. There are four division for college sports. Division I is the top division which has the best players.

(1) Division III teams are not allowed to offer sports scholarships. The students are (largely) real students, not students who spend the 90% of their time in practice and take easy or non-existent classes. At such a school, especially if it's a small school, it makes sense to be a fan because the people on the team really have a connection to you.

(2) For Division I or II there is a large divide between the players and the other college students. Hence, at such a college, rooting for the team might make less sense. However, they

are usually very good teams. So maybe, on that basis, you do want to root, root, root for the home team!

In the Olympics, one usually roots for their own country. What if (say) America offered the entire German Soccer team the opportunity to come to America, become citizens, and play for America in the Olympics, and they win? Would Americans be proud of that or feel that's not quite right? I honestly do not know the answer.

Sometimes the criteria of who to root for conflict. In Chess (Chess? Is that a sport? Yes, since there have been articles on it in *Sports Illustrated,* which I take as the definition of being a sport. Tiddlywinks is still not a sport. Oh well.) If Bobby Fischer (AMERICAN — USA! USA! USA!) played Gary Kasparov (RUSSIAN — BOO! BOO! BOO!) I, as an American, would of course root for Fischer. But wait a minute! Bobby Fischer was an awful human being and a terrible anti-semite (actually he was good at being anti-semitic). Gary Kasparov is by all accounts a nice guy and is actually against the Putin regime. So who should I root for? I would root for Kasparov.

I was going to use Boris Spassky, who Bobby Fischer really did play in 1972, but then found out that in 2005, to quote Wikipedia, *he [Spassky] signed a petition suggesting that all Jewish organizations that functioned in Russia according to the Shulchan Aruch codes should be shut down for extremism, warning about a "hidden campaign" of genocide against the Russian people and their traditional society values. Spassky later called his signature a mistake.* Since he said the signing was a mistake I suppose Spassky is not as anti-semitic as Fischer; however, to use the criteria *I always root for the least anti-semitic player* is just too depressing for words.

Is there a logic to who a fan roots for or not? Is there a logic to being a sports fan?

Chapter 3

Dispelling the Myth That the Early Logicians Were a Few Axioms Short of a Complete Set

Prior Knowledge Needed: None.

3.1 Point

There is a notion that logicians who worked in foundations early on in the field were crazy. I give examples of where this has been said and then I look at the real evidence.

(1) In Rudy Rucker's post about Alan Turing [Rucker (2006)] he says

> *it really does seem possible that Turing killed himself. Like the other logicians Gödeland Cantor, he seems to have been somewhat nuts. Funny how many logicians are crazy and irrational. A paradox.*

(2) In *Logicomix* [Doxladis and Papadimitriou (2015)], a great comic book about the foundations of logic, there is an allusion to Logicians being crazy.

(3) In Gian-Carlo Rota's book *Indiscrete Thoughts* ([Rota (2009)]) he writes:

> *it cannot be a complete coincidence that several outstanding logicians of the 20th century found shelter in asylums at some point in their lives: Cantor, Zermelo, Gödel, and Post are some.*

I've also seen two explanations for this alleged phenomena. Logicians were searching for absolute certainly and either:

(1) it drove them crazy, or
(2) thinking you can find absolute certainly means you were crazy in the first place.

So the people above, and others, give some examples of logicians being crazy and then claim that many logicians are crazy. I am reminded of people who say:

> It was cold the other day, looks like Global warming is wrong!

On Richard Zach's blog post on Logic and Madness [Zach (2009)] he debunked the notion by showing that, for Cantor and others, their craziness was unconnected to their logic. For example, Cantor was likely bi-polar. I take a different approach: looking at the the *actual record.* I will look at all of the logicians in Wikipedia's list of logicians who were born between 1845 and 1912 (1845 is when Cantor was born, 1912 is when Turing was born). I ruled out a few people who were really philosophers, and also Banach, who I don't think would call himself a logician.

For each logician on the list who meets my criteria, I say if I think they are *sane* or a *few axioms short of a complete set.* Two caveats (1) I am not a historian (2) crazy is not well defined.

You may well disagree with what years I pick and my opinions. The point is to get an intelligent discussion going.

3.2 A Set of Logicians

(1) **Wilhelm Ackerman** (1896–1962): He defined the function that bears his name. He also worked on the ϵ-calculus which formed the basis for Bourbaki's logic. Reading Bourbaki might drive one crazy; however, forming the basis for

it does not. Sane! (Ackerman that is — Bourbaki had multiple personality disorder.)

(2) **Alice Ambrose** (1906–2001): She had the longest lifespan of anyone on this list. She studied with Moore and Wittgenstein and got two PhDs. (In those days a woman had to do twice as much as a man to get a job.) She was more on the philosophy side of logic, but certainly had math training. She wrote a textbook with her husband, known as Ambrose and Lazerowitz. Sane!

(3) **Paul Bernays** (1888–1977): He worked with Hilbert on alternative set theories. Sane!

(4) **Evert Willem Beth** (1908–1964): He helped to establish Logic as a discipline. Sane!

(5) **Luitzen Egbertus Jan Brouwer** (1881–1966): (Commonly referred to as **L.E.J. Brouwer** or by his street name **Notorious L.E.J.**) He thought that all math should be constructive. This point of view lost the battle of ideas; however, that does not make him crazy. The Wikipedia article quotes Martin Davis as saying: *he felt more and more isolated, and spent his last years under the spell of totally unfounded financial worries and a paranoid fear of bankruptcy, persecution, and illness.* However, Dirk van Dalen [van Dalen (2013)], who wrote a scholarly two-volume biography of Brouwer, disagrees. I agree that Notorious is sane!

(6) **Georg Cantor** (1845–1918): He had a new way of looking at infinity that was brilliant and is now accepted. That does not make him crazy. He was also convinced that Bacon wrote the plays of Shakespeare and that Joseph of Arimathea was the father of Jesus Christ. That does not make him crazy. However, these views are indicative of a larger problem. He was often in sanitariums. Current historians think he was bi-polar.

I went to the web to see if there is a group of people (likely quite small) who thought Joseph of Arimathea was the father of Jesus Christ. No. Perhaps Cantor was an original thinker in theology as well as math. Even so: *A few axioms short of a complete set.*

(7) **Rudolph Carnap** (1891–1970): I originally thought he was more of a philosopher; however, he published in thermodynamics and the foundations of probability. He fled Hitler's regime and later refused to sign a loyalty oath in America (during the McCarthy Era). His second wife committed suicide. He led an interesting life but was sane.

(8) **Alonzo Church** (1903–1995): He invented (discovered?) the Lambda Calculus, proved that Peano Arithmetic was undecidable, and articulated what is now called the Church-Turing Thesis. These are all sane things to do. Sane!

(9) **Haskell Curry** (1900–1982): He worked in combinatory logic. There is a programming logic named after his first name! Sane!

(10) **Adolf Fraenkel** (1891–1965): The F in ZF-set-theory. Provably sane!

(11) **Gottlob Frege** (1848–1925): He hated Jews, Catholics, and the French. That might make him unpleasant to hang around, especially if you are a French Jew who converts to Catholicism; however, that does not make him crazy. He is often given as an example of someone who was crazy, but he was just a man of his time. Sad, but sane!

(12) **Gerhard Gentzen** (1909–1945): He made the cut — Sane!

(13) **Kurt Gödel** (1906–1978): He stopped eating because he thought people were trying to poison his food. They weren't. *A few axioms short of a complete set.*

(14) **Jean van Heijenoort** (1912–1986): Best known in Logic

for writing *From Frege to Gödel*. A history of Logic from . . . Frege to Gödel (duh). Best known outside of logic for being Trotsky's secretary and later a historian of that movement. His second biggest achievement was teaching Clyde logic in college. He was killed by his estranged fourth spouse. An interesting life, an interesting death, but sane!

(15) **Jacques Herbrand** (1908–1931): Has the shortest lifespan of anyone on this list (he died at 23 in a mountaineering accident). He worked in proof theory. Sane!

(16) **Arend Heyting** (1898–1980): He continued Brouwer's work on intuitionism. Sane!

(17) **David Hilbert** (1862–1943): In *Logicomix* they claim that Hilbert's son Franz had a mental illness and Hilbert cut off all contact with him. However, a review of *Logicomix* by Paolo Mancosu [Mancosu (2011)] refutes this and claims that Hilbert's son was only put away for three years and then re-joined his family. One may question if David Hilbert deserves a *World's Greatest Father* mug, but one cannot question his sanity.

(18) **Clarence Irving** (1883–1964): He took exception to Russell and Whitehead's *Principia Mathematica*'s use of material implication. I'm impressed that he read and understood Principia enough to have objections. Sane!

(19) **Stanislaw Jaskowski** (1906–1965): He worked in Intuitionistic Logics. Since I can't prove that he was crazy I assume he was sane!

(20) **William Ernest Johnson** (1858–1931): He wrote three volumes on logic which showed technical expertise but was superseded by *Principia Mathematica*. This did not drive him crazy. Sane!

(21) **Philip Jourdain** (1879–1919): He was interested in paradoxes and formed the card version of the liar's paradox. He also worked on algebraic logic. Sane! His sister Eleanor

Jourdain claimed to have traveled through time and seen ghosts, but she was not a logician.

(22) **Stephen Kleene** (1909–1994): Kleene hierarchy, Kleene star, Kleene algebras are all named after him. He also proved the recursion theorem. Did this go to his head and make him insane? Not at all. Sane!

(23) **Christine Ladd-Franklin** (1847–1930): Her PhD was on Algebra and Logic. She faced problems being a women in a man's field but kept her sanity.

(24) **Stanislaw Lesniewski** (1886–1939): He rejected axiomatic set theory (because of Russell's paradox) and tried to obtain other formal systems to replace it. A noble effort that failed. Still, he kept his sanity.

(25) **Adolf Lindenbaum** (1904–1941): He proved Lindenbaum's Lemma: every consistent theory of predicate logic can be extended to a complete consistent theory. Like many major advances, profound at the time, easy to prove now. Sane!

(26) **Leopold Lowenheim** (1878–1957): The Lowenheim of Lowenheim-Skolem. See Skolem for more on that. A model of sanity.

(27) **Jan Lukasiewicz** (1878–1956): He invented Polish notation for arithmetic expressions. Wikipedia says he thought innovatively about traditional propositional logic. Is "innovatively" a word? My spell checker does not think so but whoever wrote his Wikipedia entry thinks so. Sane!

(28) **Saunders Mac Lane** (1909–2005): (He preferred the space between Mac and Lane.) His PhD thesis was on Logic and he also worked in Category theory. He also did lots of algebra. Sane!

(29) **Carew Arthur Meredith** (1904–1976): He worked on obtaining short axiom basis for logic systems. Sane!

(30) **John von Neumann** (1903–1957): Calling him a logician seems odd since he contributed to so many fields. Sane!

(31) **Jean Nicod** (1893–1924): Co-discovered the Sheffer Stroke from which you can do everything in prop logic. Somehow that seemed important at the time. It's not. Sane!

(32) **Pyotr Novikov** (1901–1975): He proved the word problem for groups undecidable. His son Sergei Novikov won a Fields Medal in 1970 and, more importantly, is an emeritus professor at the University of Maryland! Sane!

(33) **Giuseppe Peano** (1858–1932): His Wikipedia entry calls him the founder of Mathematical Logic and Set Theory. That seems over-the-top, but not by much. His axiom system is still the standard. Sane!

(34) **Emil Post** (1897–1954): He defined the Post Correspondence Problem (I doubt he named it that) and showed it was undecidable. In the mid 1940s he posed Post's Problem, which is to find a recursively enumerable set (now called computably enumerable) that is neither decidable nor complete. This was solved in 1956 by Friedberg and Munhnik independently. He suffered from mental illness. *A few axioms short of a complete set.*

(35) **Mojzesz Presburger** (1904–1943): Presburger proved Presburger Arithmetic was decidable. What are the odds of that!? Sane!

(36) **William Quine** (1908–2000): He was more of a philosopher; however he did do some math. At Harvard he taught Symbolic Logic every fall for 50 years. That might drive some crazy, but not him. Sane!

(37) **Frank Ramsey** (1903–1930): The paper where he proved what is now known as Ramsey Theory was titled *A Problem in Formal Logic*. This paper solved a case of the Decision Problem. He regarded himself as a logician so we shall too.

Speculation: He would be surprised at where his work lead to (combinatorics) and then pleased that it lead back to logic again: *The Large Ramsey Theorem* and much work in the *Reverse Mathematics of Ramsey's Theorem.*

(38) **Raphael Robinson** (1911–1995): He worked in Logic and Number Theory. He is probably best known for his work on tiling the plane. He married Julia Robinson neé Bowman who was also a logician but born in 1919 — a little too late to be on this list. Having two academics in the same area get married might drive some crazy, but not them. Sane!

(39) **J. Barkley Rosser** (Commonly called J. Barkley Rosser.) (1907–1989): He strengthened Gödel's incompleteness theorem. Sane!

(40) **Bertrand Russell** (1872–1970): He was obsessed with the quest for certainty; however, that does not make him crazy. He had several wives (not at the same time) and believed in open marriage. He was not crazy, just ahead of his time. Sane!

(41) **Moses Schonfinkel** (1889–1942): He worked in Combinatory Logic. By 1927 he was in a sanitarium. Details are hard to come by but I'll say *a few axioms short of a complete set.* I could be wrong — see the entry on Zermelo.

(42) **Thoralf Skolem** (1887–1963): He is best known for the Lowenheim-Skolem theorem: Any consistent set of axioms has a countable model. One corollary: there is a countable model of the reals. Thinking about that might drive some crazy, but not him. Sane!

(43) **Alfred Tarski** (1901–1983): The Banach-Tarski paradox is crazy; however, Tarski was not. Sane!

(44) **Alan Turing** (1912–1954): He defined Turing Machines, though he didn't call them that. The story I had assumed was true is that the British Government made him take drugs to cure him of his homosexuality, and this drove him

to suicide. But the story doesn't quite work with the time-line. He committed suicide a few years after he stopped taking the drugs. Delayed reaction? Suicide for some other reason? Really was an accident? (That would be really odd as part of the legend of Turing is his suicide.) Turing was eccentric, though not as eccentric as portrayed in *The Imitation Game*. However, that is a far cry from insanity. In any case, since his possible suicide is the only evidence that he was crazy I say, sane!

(45) **Nicolai Vasilev** (1880–1940): The originator of non-Aristotelian logics. Sane!

(46) **Alfred North Whitehead** (1861–1947): In Russell-Whitehead's *Principia Mathematica* they spend 300 pages proving that 1+1=2. This might drive some insane but not him. Sane!

(47) **Ludwig Wittgenstein** (1889–1951): He gave away all his money and seemed to be a self-hating Jew. Odd, yes, but I originally thought he was sane. When I first posted this, Scott Aaronson left comments (see next section) that make me conclude that he was *a few axioms short of a complete set.*

(48) **Ernest Zermelo** (1871–1953): The Z in ZF set theory. He disapproved of Hitler's Regime. Hardly crazy. Rota says that Zermelo was crazy but this is a misunderstanding. Zermelo did spend time in a hospital for lung problems. The hospital was called a sanitarium, which may have confused Rota. Sane!

So what to make of all of this? Cantor, Gödel, Post, Schonfinkel, and Wittgenstein were crazy. So we have 5 out of 48 possibly clinically-certified crazy. That's around 10%. Around 6% of all people are crazy (this stat is from before the Trump election so it may be higher now). So 10% seems high, but the sample space is pretty small. Conclusion: the notion that people in logic are crazy is not well founded. In addition, the problems the five had seem unrelated to their study of logic.

However, Rota said that so many *outstanding logicians* were crazy. Since three of the five who I say were crazy were outstanding (Cantor, Gödel, Post) there may be a point here. One could look at who on my list was outstanding and see what percent of those logicians were crazy.

How to find out who in the list is outstanding? I googled

World's Greatest Logicians

and it came back with

Did you mean World's Greatest Magicians?

When I got Google to work it came back with a list of *famous logicians* which is *not* the same as *outstanding logicians*. The list seemed odd to me in that some people on it I had never heard of and I don't think are famous. Here it is in alphabetical order by last name.

(1) Pierre Abelard. A medieval French philosopher, theologian, and logician. I had never heard of him. Outside my time frame.

(2) Thomas Aquinas. Impossible to compare to modern logicians. Obviously not in my time frame.

(3) George Boole. Outstanding, but just a few years before my time frame.

(4) Lewis Carroll. A good mathematician but not an outstanding logician. Also out of my time frame. His presence on the list is a linguistic trick since he's a *famous logician* but not famous for being a logician. Ray Smullyan would love it!

(5) Haskell Curry. I didn't know he was that famous.

(6) Solomon Feferman. Outstanding, but outside of my time frame. He has the opposite problem of Frege — came along

too late to make profound discoveries on the order of, say, Gödel's incompleteness theorem.

(7) Gottlob Frege. Outstanding logician, but was around too early to do seriously hard technical work.

(8) Kurt Gödel. Outstanding logician.

(9) Saul Kripke. Outstanding, but not in my time frame.

(10) H. Dugald Macpherson. I have never heard of him. Either he shouldn't be on the list or I am not as up on famous logicians as I thought. Not in my time frame.

(11) Bertrand Russell. More famous as a philosopher.

(12) Raymond Smullyan. My personal hero as he straddles recreational and serious math and makes a mockery of the distinction. Not outstanding and not in my time frame.

(13) Alfred Tarski. Outstanding logician.

(14) Alan Turing. Outstanding logician.

(15) Alfred North Whitehead. More famous as a philosopher.

Here is my (debatable) list of who was outstanding:

Brouwer (sane), Cantor (crazy), Church (sane), Frege (sane), Gödel (crazy), Hilbert (sane), Kleene (sane), von Neumann (sane), Peano (sane), Post (crazy), Tarski (sane), Turing (sane). Of the twelve, three are crazy. That's 25%, which is a lot. But the set consists of only twelve people, which is hardly a good sample size. This needs more study. Give me a grant and I'll study it.

Was it crazy to spend so much time and effort on this one section? I am not on the logic list, nor was I born between 1845 and 1912 so the answer is not relevant to the study.

3.3 Some Comments on Section 3.2

In response to my original blog post, John Sidles commented that in Physics, statistical mechanics is associated with madness

*In physics it is statistical mechanics that is associated*ᐧ *with madness, as vividly depicted in the opening sentence of David Goodstein's classic textbook States of Matter.*

Ludwig Boltzman, who spent much of his life studying statistical mechanics, died in 1906, by his own hand. Paul Ehrenfest, carrying on the work, died similarly in 1933. Now it is our turn to study statistical mechanics. Perhaps we will be wise to approach the subject cautiously.

Of course nowadays no one should take the association seriously... or *should* we? :):):)

I leave it to some Physics blogger to do a detailed study of their myth.

In response to the same post, Xamauel commented about Lewis Carroll:

Though well outside the window you're looking at, Charles Lutwidge Dodgson is an interesting example: among non-mathematicians, he's probably more well-known (as "Lewis Carroll") than all of the mathematicians on your list *combined*. What's more, because of Alice's Adventures in Wonderland, people generally assume he was a regular hatter, but he actually was not. Anyway, I wouldn't be surprised if he contributes something to the stereotype.

"Is 'innovatively' a word?" Books.google.com says "Yes!"

The fact that Lewis Carroll is better known than Kurt Gödel is sad but true.

Gowers noted that I left off whether or not Ramsey was sane. I give his entire comment:

I see you don't tell us whether Ramsey was sane or a few axioms short of a complete set. This coded message has led me to your short proof that P *is not equal to* NP, *which not many people realize is being kept secret by an elite group of theoretical computer scientists. I have been forced into hiding with this dangerous information.*

On the one hand, I am not sure how he reached that conclusion. On the other hand, I make it a policy to never argue with a Field's medal winner.

Someone named Anonymous had the following comment:

I wouldn't say Gödel was crazy. Having a false belief does not make one crazy, if it did we should call large segments of the US population crazy because of not just false, but provably stupid, beliefs.

A person having a mental disorder is not necessarily crazy. But why is it popular to make claims like that most logicians are crazy? Because it looks paradoxical. Most people don't understand what logicians have done and it looks strange to them. It is also entertaining, like gossips about the private lives of movie stars or politicians, or if you are in the UK (or other un-united ones) about royals. If you ask me, people interested in following these gossips or watching those silly evening shows on TV are way crazier than any logician you have listed above as crazy or a few axioms short of [a complete set]. But it is a democracy, so Socrates is sentenced to death.

There is a lot to unpack here; however, I make one comment about the comment. Anon is right that people gossip about things that seem paradoxical:

(1) It is commonly believed that Einstein, one of the 20th century's great geniuses, did poorly in school. What an irony! The story is also hopeful: *I did poorly in school, just like Einstein. So people who tell me "you're no Einstein" are wrong!* Alas, the story, while well circulated, is not true. Einstein did quite well in school. But people like to believe it since it is paradoxical.

(2) For some comedians, say C, there have been documentaries titled: *Behind the laughter: The true story of C.* In such documentaries they point out how miserable C's life was. But this is a skewed sample — if a comedian has a happy life they don't bother making a story about it. People like to believe that comedians are unhappy since it's paradoxical.

Richard Elwes comments on Cantor:

> *Thank you for busting the myth.*
> *I think Cantor is the original source of the stereotype.*
> *After all, he was the first to stare infinity in the eye.*
> *And it drove him mad, I tell you,* **mad!**

The original post classified Wittgenstein as sane. Fellow blogger Scott Aaronson asked me to reconsider, which I did, as you can see above. Here is his comment in full.

> *Bill, you might want to consider your "sane" classification for Wittgenstein!*
>
> *According to Ray Monk's biography, during World War I, he volunteered for the Austria-Hungarian army and repeatedly begged to be sent to wherever the heaviest fighting was, not because he cared at all about the war's outcome, but because he thought the experience of battle would ennoble and purify him.*
>
> *One of his main influences was Otto Weininger, a raving misogynist and anti-semite who shot himself;*

Wittgenstein was ashamed that he didn't kill himself as well.

Wittgenstein abandoned philosophy to become an elementary-school teacher in a small village in Austria, where he quickly became notorious for boxing boys' ears and pulling girls' hair when they couldn't solve a math problem (he later went door-to-door to beg the parents' forgiveness).

My impression, after reading Monk, was that Wittgenstein was not merely insane but committed to insanity as a philosophical ideal.

Some commenters disagreed that this makes Wittgenstein crazy. However, I side with Scott and declare him (Wittgenstein that is, not Scott) to be a few limit points short of a Banach space.

Ron Fagin leaves a comment about Church which indicates Church was eccentric, perhaps very eccentric, but not crazy.

You quote Rota as saying:
"*It cannot be a complete coincidence that several outstanding logicians of the 20th century found shelter in asylums at some point in their lives: Cantor, Zermelo, Gödel, and Post are some.*"

However, you do not quote the next sentence in that article, which is about Alonzo Church:
"*Alonzo Church was one of the saner among them, though in some ways his behavior must be classified as strange, even by mathematicians' standards.*"

Rota goes on to tell stories about Church, including the following:
"*Every lecture began with a ten-minute ceremony of erasing the blackboard until it was absolutely*

spotless. We tried to save him the effort by erasing the board before his arrival, but to no avail. The ritual could not be disposed of; often it required water, soap, brush, and was followed by another ten minutes of total silence while the board was drying. Perhaps he was preparing the lecture while erasing; I don't think so. His lectures hardly needed any preparation. They were a literal repetition of the typewritten text he had written over twenty years ago, a copy of which was to be found upstairs in the Fine Hall library."

The comments on Wittgenstein and Church remind one that it may be hard to distinguish crazy from eccentric. As such, my classifications can be debated. Nevertheless, the point is made that not very many logicians are crazy; however, statistical mechanics...

References

Doxladis, A. and Papadimitriou, C. (2015). *Logicomix* (Bloomsbury), edited by Fabrizio Palombi.

Mancosu, P. (2011). Logic and madness, *The Journal of Humanistic Mathematics*, pp. 137–152 https://scholarship.claremont.edu/jhm/vol1/iss1/10/.

Rota, G.-C. (2009). *Indiscrete thoughts* (Birkhauser, Boston), edited by Fabrizio Palombi.

Rucker, R. (2006). Alan turing, http://www.rudyrucker.com/blog/2006/10/02/alan-turing/.

van Dalen, D. (2013). *L.E.J. Brouwer — Topologist, intuitionist, philosopher* (Springer, New York, Heidelberg, Berlin), e-book, third edition.

Zach, R. (2009). Logic and madness, http://www.ucalgary.ca/rzach/blog/2009/09/logic-and-madness.html.

Chapter 4

What is a Simple Function?

Prior Knowledge Needed: None.

4.1 The Question

In my sophomore discrete math course I asked the following question: For each of the following sequences find a simple function $A(n)$ such that the sequence is $A(1)$, $A(2)$, $A(3)$, ...

(1) 10, -17, 24, -31, 38, -45, 52, $\cdots$
(2) -1, 1, 5, 13, 29, 61, 125, $\cdots$
(3) 6, 9, 14, 21, 30, 41, 54, $\cdots$

These are not trick questions. The answers, and the point I am trying to make, are on the next page.

4.2   The Answer and the Point

I give the answers and also how one might derive them.

(1) 10, -17, 24, -31, 38, -45, 52, $\cdots$
We can take care of the alternating positive and negative by using $(-1)^{n+1}$. Hence we can look at
10, 17, 24, 31, 38, 45, 52, $\cdots$
Notice that each term is 7 more than the following one. Hence the pattern is of the form $7n + b$. We easily find that $b = 3$. Hence
$$A(n) = (-1)^{n+1}(7n + 3).$$

(2) -1, 1, 5, 13, 29, 61, 125, $\cdots$
A standard technique is to look at the differences between consecutive elements:
$1 - (-1) = 2$
$5 - 1 = 4$
$13 - 5 = 8$
$29 - 13 = 16$
Aha! These are powers of two! We guess that the sequence is of the form $2^n + b$ and find $b = -3$. Hence
$$A(n) = 2^n - 3.$$

(3) 6, 9, 14, 21, 30, 41, 54, $\cdots$
Again, look at the differences between consecutive elements:
$9 - 6 = 3$
$14 - 9 = 5$
$21 - 14 = 7$
Looks like the difference is linear. Hence, the original sequence is quadratic. By guessing $an^2 + bn + c$ and interpolating we obtain
$$A(n) = n^2 + 5.$$

I had originally thought that all three sequences would be fairly easy to recognize. The students found Question 1 to be easy, but Questions 2 and 3 to be difficult. In retrospect they are right. The patterns are not easy to see, and I had not taught them the difference technique.

Some noted correctly that the term *simple function* was not defined in class. I meant a function that does not involve summations or recurrences, though I do not think that quite captures the notion. For example, a polynomial that interpolated the values of a sequence is not what I intended for Problem 2. I could ask for the function that fits the sequence whose description has the smallest Kolmogorov complexity of all such functions; however, that's not quite right for a sophomore-level discrete math class where the students struggle to learn induction. (See Chapter 24 for information on Kolmogorov Complexity.)

One student, in earnest, emailed me the Wikipedia entry on *simple functions*, which defines them as *a linear combination of indicator functions of measurable sets.* Hmmm — not quite what I had in mind for this class. The student wanted to know if he could use it. I told him no; however, in retrospect, I wonder what he would have come up with.

I did not know that simple function had a formal definition. Perhaps I should have since "simple sets" have a formal definition in computability theory: A *simple set* is a co-infinite set that is computably enumerable (which used to be called recursively enumerable), such that every infinite subset of the complement is not computably enumerable. Clyde notes that only a logician would call this simple.

I was once fooled by another term whose meaning in Mathematics is different than its meaning in English. Andrews and Baxter [Andrews and Baxter (1989)] have a paper titled: *A Motivated Proof of the Rogers-Ramanujan Identities.* I went to a talk on this paper thinking *Great! I will get a proof that I have a sense of why it works the way it does!* Sadly no. It turns out that *motivated* is a technical term. The proof was not motivated in the English sense of the word.

While looking up these references to write this chapter I came across a paper by Bressoud [Bressoud (1983)] titled: *An Easy Proof of the Rogers-Ramanujan Identities.* I hope that *easy* has retained its English meaning.

4.3   Comments on Section 4.2

The original point of the post was that normal English words and phrases can be interpreted in unexpected ways. This point is neither surprising nor controversial. Maybe because of this, the discussion went off topic: Many of the commenters objected to the question I was asking. Why? There were two kinds of objections.

(1) *The question is ambiguous, since there is no unique answer.*
(2) *Such questions have little pedagogical value.*

Here is a typical (anonymous) comment that touches on both points:

> *Even if this kind of questions* [sic] *made sense from [a Machine Learning] point of view, the size of data set is too small. Kolmogorov complexity neither makes sense, it completely depends on the the K function you have in mind. I strongly dislike these* [sic] *kind of question, you are essentially asking students to guess what you have in mind. Do you think it is going to help them in predicting the stock market or something? What is the point of these questions?*

The On-Line Encyclopedia of Integer Sequences (OEIS) contains all integer sequences (well, not all, but lots). Here is an explicit (anonymous) comment related to the second point:

> *All of them can be looked up in OEIS. Meaningless exercise.*

Here is a paraphrase of how I responded to these comments and others like them:

Response to the Earlier Commenter: It is true that any finite sequence can be extended any way one wants, and one can find some polynomial that gives those values. That's why I said to *give a simple function.* On this level I think it is okay to give easy sequences (easier than I gave) and ask the students to guess what's next. I'll also point out that these are not trick questions. I am not going to say $2, 4, 6, 8$ — *the next term is 100* and then point to some polynomial of degree four that gives those values. Also, you took my *Kolmogorov* reference too seriously.

Response to the Later Commenter: You are raising an interesting point: *Now that we have computers that can do BLAH better than humans, we can stop training humans to do BLAH.* Should elementary school children still have to learn the multiplication tables now that we have calculators? I think that, for small numbers, yes. Should college students be able to recognize patterns? I think that, on the level of fairly easy sequences (easier than I gave), the answer is yes.

I end with a concise (anonymous) comment from the post that supports my point of view. And no, it is not my own comment.

> *I find it amazing when people complain that such problems do not have a unique solution. It seems like everything in life should have a unique solution... Noticing patterns is important because most people work by observing, conjecturing and proving.*

CLYDE'S COMMENTS:
First, I basically agree with what Bill has said. Recognizing patterns is a useful skill in general science, computer science,

computer programming, life in general, etc. My minor dis-
agreement is the suggestion that these particular problems are
too difficult. (Although, when I ask similar homework questions, I do give the hint to look at differences.) Some students
can do such problems, others cannot. For example, they may
not realize that when the differences are linear, the function itself is quadratic. That is what homework is for. The next time
they see something similar they will be better prepared. Even
if they are never able to solve such problems, the process of
thinking about and struggling with them is a valuable learning
experience.

Consider the list $2, 4, 6, 8$. I will go out on a limb and say
that almost everyone reading this will immediately think of
10 as the next value. I am sure that the simplest sequence
is the (positive) even numbers in order; I do not know how
to justify it formally. I do not care that the sequence is too
short for Kolmogorov complexity to apply. If someone says it
is some other value, I will not argue — I have learned that such
arguments are futile — but I believe that person is misguided.

I once had a student who came to see me for help with
a homework problem. As part of the problem, the student
needed to find a particular function, whose first four values
turned out to be $1, 2, 4, 8$. The student had no idea what the
next value should be or even might be. It is possible that 16 is
not the next value for that particular function, but at least it
should be an obvious possibility. In my experience, the ability
to recognize such sequences strongly correlates with the ability
to program.

There are many reasons to ask such questions — too many
to discuss here. One reason is to make mathematics more real
and interesting, so students do not think of mathematics as
merely being given an equation and then solving it. One type

of question is to start with an actual problem where you have to derive a function, generate the first few values, guess the formula, and finally prove the formula.

For example, assume that you want to know how many comparisons bubble sort uses in the worst case. We can count the number of comparisons explicitly for $n = 1, 2, 3, 4, 5$ getting $0, 1, 3, 6, 10$, respectively. Looking at differences, it is easy to guess that the formula is $n(n-1)/2$. Given the guess, it is easy to prove the formula. Yes, this particular problem is easy to solve without guessing the formula, but many problems are not so easy. I try to make at least some homework problems have this flavor.

4.4 A Nonpattern: Regions in a Circle

There are many patterns where the first obvious guess is wrong. That is one reason it is important to prove that pattern-fitting data is correct. Here is a famous example: Consider the maximum possible number of regions formed by connecting every pair of n points on a circle. Here are the cases for $n = 1, 2, 3, 4, 5$:

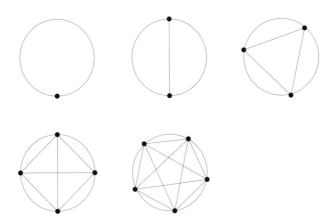

Fig. 4.1   Regions formed by connecting points on a circle for $n = 1, 2, 3, 4, 5$.

The number of regions are 1,2,4,8,16, respectively. It is tempting to guess that the number of regions is 2^{n-1}. It turns out that, for $n = 6$, the number of regions is only 31. (Note that you cannot place the six points *evenly* around the circle, because three chords will meet in the center reducing the number of regions.) The actual number of regions for n points is

$$\binom{n}{4} + \binom{n}{2} + 1$$

We give three proofs. All three proofs use the following lemma:

Lemma 4.1. *Assume there are n points on a circle with a chord drawn between every pair of points. Assume that no three chords intersect (at the same point). Then*

(1) The number of chords in the circle is $\binom{n}{2}$.
(2) The number of intersection points is $\binom{n}{4}$.

Proof.
1) This is obvious.
2) Each set of four points on the circle produces exactly one intersection point (with its six chords). So the number of intersection points is $\binom{n}{4}$.   □

Let n be the number of points on the circle. To create the maximum number of regions, every intersection point consists of only two chords. If more than two chords intersect at a point, just perturb the endpoints of one chord.

4.4.1   *Direct Proof*

Assume that some chords have already been drawn, and we start drawing a new chord from point a to point b on the circle. When the new chord creates its first intersection point by crossing an existing chord it splits a region in two. When the new chord creates its second intersection point by crossing a second

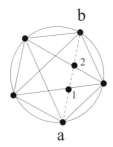

Fig. 4.2   New regions formed by creating a chord connecting a to b.

existing chord it splits another region in two. It continues in this fashion until it reaches point b. Each chord it crosses adds one more intersection point and one more region. Figure 4.2 gives an example as the last edge is added for $n = 5$. Two new intersection points (in bold, numbered 1 and 2) are added, and three regions are split in two.

So the total number of regions is the number of intersection points, plus the number of chords, plus one for the original region (which starts with no chords). The result now follows from Lemma 4.1.

4.4.2   *Proof using Euler's polyhedron formula*

Euler's polyhedron formula
$$V + F - E = 2$$
can be used to avoid the original argument about how adding a new chord creates regions. Let R be the number of regions, I be the number of intersection points, and C be the number of chords. In our case, the number of faces is one more than the the number of regions since faces includes the outside of the circle, so $F = R + 1$. The number of vertices is the number of intersection points plus the number of points on the circle, so $V = I + n$. Each chord is initially an edge, and each intersection point creates two new edges (one on each chord). Also, each arc of the circle is an edge. So $E = 2I + C + n$.

Substituting into Euler's polyhedron formula gives

$$(I + n) + (R + 1) - (2I + C + n) = 2$$

Simplifying gives

$$R = I + C + 1$$

The rest follows by Lemma 4.1.

4.4.3   *Proof by Mathematical Induction*

Once we know the formula, it is also easy to prove it by induction. Assume that $n - 1$ points create

$$\binom{n-1}{4} + \binom{n-1}{2} + 1$$

regions. By Lemma 4.1 one more point adds $n - 1$ new cords and $\binom{n}{4} - \binom{n-1}{4}$ new intersection points. So the number of new regions is

$$\left(\binom{n}{4} - \binom{n-1}{4} \right) + (n - 1)$$

The total number of regions is therefore

$$\left[\binom{n-1}{4} + \binom{n-1}{2} + 1 \right] + \left[\left(\binom{n}{4} - \binom{n-1}{4} \right) + (n - 1) \right]$$

$$= \binom{n}{4} + \left(\binom{n-1}{2} + (n - 1) \right) + 1$$

$$= \binom{n}{4} + \binom{n}{2} + 1$$

4.5   More Nonpatterns: Borwein Integrals

The following two sequences are called Borwein Integrals, named after David and Jonathan Borwein [Borwein and Borwein (2001)], the father-and-son mathematicians who first presented and explained them. For another explanation see the paper of Hanspeter Schmid [Schmid (2014)].

$$\int_0^\infty \frac{\sin x}{x} = \frac{\pi}{2}$$

$$\int_0^\infty \frac{\sin x}{x} \frac{\sin \frac{x}{3}}{\frac{x}{3}} = \frac{\pi}{2}$$

$$\int_0^\infty \frac{\sin x}{x} \frac{\sin \frac{x}{3}}{\frac{x}{3}} \frac{\sin \frac{x}{5}}{\frac{x}{5}} = \frac{\pi}{2}$$

$$\int_0^\infty \frac{\sin x}{x} \frac{\sin \frac{x}{3}}{\frac{x}{3}} \frac{\sin \frac{x}{5}}{\frac{x}{5}} \frac{\sin \frac{x}{7}}{\frac{x}{7}} = \frac{\pi}{2}$$

$$\int_0^\infty \frac{\sin x}{x} \frac{\sin \frac{x}{3}}{\frac{x}{3}} \frac{\sin \frac{x}{5}}{\frac{x}{5}} \frac{\sin \frac{x}{7}}{\frac{x}{7}} \frac{\sin \frac{x}{9}}{\frac{x}{9}} = \frac{\pi}{2}$$

$$\int_0^\infty \frac{\sin x}{x} \frac{\sin \frac{x}{3}}{\frac{x}{3}} \frac{\sin \frac{x}{5}}{\frac{x}{5}} \frac{\sin \frac{x}{7}}{\frac{x}{7}} \frac{\sin \frac{x}{9}}{\frac{x}{9}} \frac{\sin \frac{x}{11}}{\frac{x}{11}} = \frac{\pi}{2}$$

$$\int_0^\infty \frac{\sin x}{x} \frac{\sin \frac{x}{3}}{\frac{x}{3}} \frac{\sin \frac{x}{5}}{\frac{x}{5}} \frac{\sin \frac{x}{7}}{\frac{x}{7}} \frac{\sin \frac{x}{9}}{\frac{x}{9}} \frac{\sin \frac{x}{11}}{\frac{x}{11}} \frac{\sin \frac{x}{13}}{\frac{x}{13}} = \frac{\pi}{2}$$

$$\int_0^\infty \frac{\sin x}{x} \frac{\sin \frac{x}{3}}{\frac{x}{3}} \frac{\sin \frac{x}{5}}{\frac{x}{5}} \frac{\sin \frac{x}{7}}{\frac{x}{7}} \frac{\sin \frac{x}{9}}{\frac{x}{9}} \frac{\sin \frac{x}{11}}{\frac{x}{11}} \frac{\sin \frac{x}{13}}{\frac{x}{13}} \frac{\sin \frac{x}{15}}{\frac{x}{15}} =$$

$$\frac{467807924713440738696537864469\pi}{935615849440640907310521750000} \sim 0.9999999999852937186 \times \frac{\pi}{2}$$

Why the breakdown at 15? It is related to the fact that

$$\frac{1}{3} + \frac{1}{5} + \cdots + \frac{1}{13} < 1$$

but

$$\frac{1}{3} + \frac{1}{5} + \cdots + \frac{1}{15} > 1.$$

Is there an even longer sequence that seems to have a pattern but does not? Yes! Also by the Borweins:

$$\int_0^\infty 2\cos(x)\frac{\sin x}{x} = \frac{\pi}{2}$$

$$\int_0^\infty 2\cos(x)\frac{\sin x}{x}\frac{\sin\frac{x}{3}}{\frac{x}{3}} = \frac{\pi}{2}$$

$$\vdots$$

$$\int_0^\infty 2\cos(x)\frac{\sin x}{x}\cdots\frac{\sin\frac{x}{111}}{\frac{x}{111}} = \frac{\pi}{2}$$

but

$$\int_0^\infty 2\cos(x)\frac{\sin x}{x}\cdots\frac{\sin\frac{x}{113}}{\frac{x}{113}} < \frac{\pi}{2}$$

Why the breakdown at 113? It is related to the fact that

$$\frac{1}{3}+\frac{1}{5}+\cdots+\frac{1}{111} < 2$$

but

$$\frac{1}{3}+\frac{1}{5}+\cdots+\frac{1}{113} > 2.$$

I have not been able to find an extension of these results that would break down when

$$\frac{1}{3}+\frac{1}{5}+\cdots+\frac{1}{2n+1} > 3.$$

References

Andrews, G. and Baxter, R. (1989). A motivated proof of the Rogers-Ramanujan identities, *The American Mathematical Monthly*, pp. 401–409.

Borwein, D. and Borwein, J. (2001). Some remarkable properties of sinc and related integrals, *Ramanjuan Journal*, 1, pp. 73–89, http://docserver.carma.newcastle.edu.au/249/2/142-Borwein-Borwein.pdf.

Bressoud, D. (1983). An easy proof of the Rogers-Ramanujan identities, *Journal of Number Theory*, pp. 235–241.

Schmid, H. (2014). Two curious integrals and a graphic proof, *ELEM* **69**, pp. 11–17.

Chapter 5

How Do Mathematical Objects Get Named?

Prior Knowledge Needed: None.

5.1 Point

There are many different ways Mathematical Objects get their names. We discuss some of these ways.

5.2 Name the Theorem After the Person Who Proved it

If a theorem is named after the person who first proved it (which can be tricky) you are giving proper credit. But some people have more than one theorem named after them. It's fine that these people get credit, but there is an ambiguity. There are seven results called *Schur's Theorem*. Issac Schur must have been an excellent mathematician to have seven theorems named after him! But wait a minute! One of them was due to a different Schur. Still, six theorems is impressive. Here are the topics of the seven theorems, the last of which was due to Axel Schur (not Issac).

- Ramsey Theory
- Making change
- Linear Algebra (often called Schur Decompositions)

- Functional Analysis
- Number Theory
- Riemannian Geometry
- Differential Geometry

To confuse matters further (1) sometimes Issac Schur's first name is spelled Issai, and (2) there is yet another mathematician with his last name: Friedrich Schur.

5.3 Name the Mathematical Object After Euler or Gauss

If you don't know what to call a mathematical object then just use Euler or Gauss, and you're probably correct.

According to Wikipedia (as of 2018) Euler has the following named after him:

(1) 2 conjectures
(2) 11 equations
(3) 7 formulas
(4) 4 functions
(5) 3 identities
(6) 9 numbers
(7) 10 theorems
(8) 2 laws

According to Wikipedia (as of 2018) Gauss has the following named after him:

(1) 5 conjectures (one is now called *the Hawaii conjecture. Really!*)
(2) 1 equation and 3 inequalities
(3) 5 formulas
(4) 4 functions
(5) 5 identities

(6) 3 numbers (includes classes of numbers like the Gaussian Integers)
(7) 17 theorems and 3 lemmas
(8) 3 laws

For both Euler and Gauss many of these math objects were co-authored.

(1) The *Gauss-Bonnet Theorem* in geometry. Gauss was aware of a version of the theorem but never published it. Bonnet proved a special case in 1848. I could not find if Gauss and Bonnet ever met or corresponded. Gauss lived from 1777–1855 and Bonnet lived from 1819–1892, so they may have.
(2) The *Euclid-Euler Theorem* on Perfect Numbers (proven in Chapter 21). Euclid is thought to have been born in the 4th century BC and died in the 3rd century BC. Euler lived in the 18th century. Suffice to say, they never met nor corresponded. They get joint credit because each proved different directions of an if-and-only-if.
(3) The Euler-Fermat Theorem. This one is easy to state: For all a, n $a^{\phi(n)} \equiv a \pmod{n}$ where $\phi(n)$ is the number of numbers in $\{1, \ldots, n\}$ that are relatively prime to n. Since Fermat lived in the 17th century they never met nor corresponded. They get joint credit since Fermat proved the case where n is prime and Euler generalized to obtain the current version.

5.4   Hope Someone Names Your Theorem After You

Martin Kruskal (Clyde's father) and Norman Zabusky rediscovered Soliton Waves, which are important in Mathematics and Physics. (They were first discovered by John Scott Russell in the early 1800s and called *Waves of Translation*.)

Clyde and his young daughter, Rebecca (hence Martin's granddaughter) had the following conversation:

Rebecca: Daddy, how come Soliton waves aren't called *Kruskal-Zabusky waves.* That would be so cool!

Clyde: Do you know what a Soliton wave is?

Rebecca: No, but it would still be cool to have something named after a Kruskal. So answer my question, why aren't they called Kruskal-Zabusky Waves?

Clyde: You can't name things after yourself.

Rebecca: Why not?

Why not indeed! An unwritten rule? Probably! And the rule makes sense. It also makes Gauss and Euler's plethora of things named after them more impressive.

However, Rebecca's dream has come true! I often search arXiv for my own papers since I know they are there. I once accidentally had it set for papers that mention me anywhere, not just as an author. Ian Morrison [Gasarch and Kruskal (1999)] named something *The Gasarch-Kruskal Theorem*!

It's about exactly when you can load dice to get fair sums. So Rebecca is now happy!

5.5 Math Objects Named Descriptively

There are many examples of concepts named in a way that seems to describe or hint at the concept. We just give two.

Girth of a graph: The *girth of a graph* is its shortest cycle. In English *girth* means the measurement around the middle, often a person's waist.

Byzantine General's Problem: The name *Byzantine General's Problem* actually refers to a set of problems where there are n processors, some faulty, and they have to agree on something (perhaps a bit, perhaps a leader). We give an example which also says how they got this name. The example and the name are both due to Leslie Lamport *et al.* [Lamport *et al.* (1982)].

The name comes from the following scenario: A set of generals each control part of the Byzantine army. They can communicate pairwise by messenger. They want to either all agree to attack a city or all agree to not attack. If a majority wants to attack, they will all attack. This avoids a half-ass effort where only (say) half of them attack. Some of the generals are traitors. Consider the following: there are 7 generals, 6 are loyal. Of those 6, 3 want to attack and 3 do not want to attack. The traitorous general tells the pro-attack forces ATTACK and the anti-attack forces DON'T ATTACK. The result is that there is an attack by only 3 forces. This is the kind of scenario you want to avoid. So, is there a protocol so that, at the end of it, all of the loyal generals know what the majority of loyal generals want to do? (Note that they still might not know who the traitor is.) Often problems of this type can be solved when more than 2/3 of the generals are loyal.

5.6 Math Objects Named For a Funny Reason

This section was the original inspiration for this chapter.

5.6.1 *The Governor's Theorem*

How much math should our public officials know? Basic probability and statistics so they can follow the arguments that their science advisers give them. Also they should hire good, objective science advisers and listen to them.

How much trigonometry should a governor know? Should a governor know the angles of a 3-4-5 triangle? The following true story is paraphrased from an article by Skip Garibaldi [Garibaldi (2008)] with the delightful title *Somewhat more than governors need to know about trigonometry*.

In June 2004, Governor Jeb Bush of Florida was giving a talk to promote state-wide annual testing of students in public schools. A high school student asked him
            What are the angles in a 3-4-5 triangle?
He responded

> *I don't know. 125, 90, and whatever is left to add up to 180.*

Note that:

(1) he knew that 3-4-5 triangle has a 90 degree angle,
(2) he knew that the angles of a triangle add up to 180, but
(3) he didn't realize that $125 + 90 > 180$.

Still, I suspect most governors would do worse. The real answer is 90, (approx) 53.1, and (approx) 36.9. A retired math professor was later quoted as saying: I would not expect many mathematicians to know that.

The paper then proves the following:

The Governor's Theorem: If a right triangle has integer side lengths then the acute angles are irrational when measured in degrees.

The name of the theorem comes, not from the author's name (no, Jeb Bush was not a coauthor), nor from the mathematical content, but from the story that inspired the theorem.

5.6.2 *Greedy Galois Games*

I know what you are thinking. Galois was a mathematician, so whatever *Greedy Galois Games* are, they must be connected to mathematics that Galois worked on. I thought so too. I had the following speculations of what a *Galois Game* would be (I didn't quite catch the *Greedy* in the title).

(1) The game has a parameter $d \in \mathbb{N}$ with $d \geq 5$. The players alternate, picking the coefficients in $\mathbb{Z}$ of a degree d polynomial. If at the end the polynomial has a solution in radicals then Player I wins, else Player II wins. Variants: (1) have the goals of the players switch, (2) have the coefficients come out of an extension of $\mathbb{Z}$ that is announced ahead of time, like the Gaussian Integers (Yeah Gauss!).

(2) The game has a parameter $d \in \mathbb{N}$ with $d \geq 2$. Let D_1, D_2 be integral domains. The players alternate, picking the coefficients in D_1 of a degree d polynomial. If at the end the polynomial has a solution in D_2 then Player I wins, else Player II wins. Variants: (1) have the goals of the players switch, (2) have restrictions on if picking 0 is allowed.

Neither of these speculations is close to what the author meant. Larry Washington, Sam Zbarsky, and I published a paper [Gasarch *et al.* (2018)] on strategies for these games. And now, back to our story.

Aside from being a mathematician, what is Galois best known for?

He died in a duel!

The paper used Galois's name *not* as a good mathematician but as a bad duelist.

Here is a Greedy Galois Game:

(1) Alice and Bob are duelists. Alice has probability p of hitting Bob. Bob has probability q of hitting Alice.

(2) Alice fires $a_1 = 1$ shots at Bob.

(3) If Bob survives than Bob fires b_1 shots at Alice. b_1 is picked to be the least number so that the probability of Bob winning during this stage exceeds that of Alice winning earlier.

(4) If Alice survives then she shoots a_2 shots at Bob. a_2 is picked to be the least number so that the probability that Alice winning during this stage exceeds that of Bob winning earlier.

(5) Etc.

The paper is about deriving $a_1, b_1, a_2, b_2, \ldots$ from p, q. It involves some interesting math.

Could they have picked a different bad duelist to name their games after? They could have picked Hamilton, but then I would think that the game had to do with Hamiltonian Graphs or Quaternions.

5.7 Names Change: Conjecture to Theorem

The following is *Baudet's conjecture:*

If ℕ *is partitioned into a finite number of classes then one of the classes has arbitrarily long arithmetic progressions.*

This statement is no longer called *Baudet's conjecture.* Why? Because van der Warden proved it. It is now called *van der Warden's Theorem.* Bad luck for Baudet. Even worse luck — the Wikipedia page on van der Warden's Theorem does not mention Baudet at all (though other web sources do).

The following is *Vázsonyi's conjecture:*

Order trees under the minor ordering. (1) there is no infinite descending sequence of trees, (2) there is no infinite antichain of trees.

This statement is no longer called *Vázsonyi's conjecture:* Why? Because Joseph Kruskal (Clyde's Uncle) proved it. It is now called *The Kruskal Tree Theorem.* Bad luck for Vázsonyi.

However, the Wikipedia page on *The Kruskal Tree Theorem* does mention Vázsonyi, so he is better remembered than Baudet.

The following is *Goldbach's conjecture:*

Every even number $n \geq 4$ can be written as the sum of two primes.

This statement is still called *Goldbach's conjecture*. Why? Because nobody has proved it yet. When someone does will Goldbach's name be as forgotten as Baudet's? The name has been on it since 1742 so it may stick around for a while after the conjecture is proven. Baudet's and Vázsonyi's conjectures were not open long enough to get iconic status.

The following is (was?) *Fermat's Last Theorem:*

For all $n \geq 3$ there is no solution in positive integers to the equation $x^n + y^n = z^n$.

Fermat wrote the statement in the margins of his copy of Diophantus's *Arithmetica* (translated to English):

It is impossible to separate a cube into two cubes, or a fourth power into two fourth powers, or in general, any power of higher than the second into two like powers. I have discovered a truly marvelous proof of this, which this margin is too narrow to contain.

Fermat did not have such a proof. In later life he proved the $n = 4$ case, which he would not have done had he had the proof. The problem for general $n \geq 3$ came to be known as *Fermat's Last Theorem*.

In 1993 Andrew Wiles proved Fermat's Last Theorem using techniques that were far beyond anything known in Fermat's day (or even in Gauss's day). To celebrate this Tom Lehrer added the following verse to his song *That's Mathematics!*

Andrew Wiles gently smiles
Does this thing, and voila!

Q.E.D, we agree
And we all shout hurrah!
As he confirms what Fermat
Jotted down in that margin
Which could've used some enlargin'.

Another celebration of the proof of Fermat's Last Theorem is the following limerick whose origin I have been unable to trace:

A challenge for many long ages
Had baffled the savants and sages
    Yet at last came the light
    Seems that Fermat was right
To the margin add 200 pages

People still refer to *Wiles's proof of Fermat's Last Theorem.* I wonder if people will just call it *Wiles's Theorem* eventually. Even so, the name Fermat will live on in Tom Lehrer's song and the anonymous limerick!

5.8   Names Change: Historical Accuracy

5.8.1   *The WFI Algorithm*

Definition 5.1. The *all-pairs-shortest-path (APSP)* problem is, given a weighted graph G, return for all pairs of vertices the length of the shortest path between them.

In 1962 Robert Floyd published an algorithm for APSP that ran in $O(n^3)$ time and was an early example of Dynamic Programming. For a time it was called *Floyd's Algorithm*. Floyd used (and acknowledged) ideas from Warshall's transitive closure paper of 1962. Bernard Roy had a similar algorithm in 1959. The algorithm has been called by the following names:

(1) Floyd's algorithm

(2) Floyd-Warshall algorithm
(3) The Roy-Warshall algorithm
(4) The Roy-Floyd algorithm
(5) The WFI algorithm (I got this from both Wikipedia and MathWorld. Neither says why it's WFI. I think it should be WFR but I don't have their reach.)

Since all three authors had contributions I prefer *the WFR algorithm*. When I tell this story to students now they ask *How do they know that Floyd didn't read the version that Roy posted?* Of course, in those days you couldn't post articles to the web. But this raises the point: it may be much harder to believe independent discovery now then it was then.

5.8.2 *The Cook-Levin Theorem*

Cook proved Cook's Theorem (SAT is NP-complete) in 1971. For a long time the statement was known as *Cook's Theorem*. It was later discovered that Levin proved it at around the same time. It is now called *The Cook-Levin Theorem*.

There is a professor in my department named Dave Levin. My students asked if he was related to Leonid Levin (of Cook-Levin). On a whim I said that Dave was Leonid's nephew. The students believed me. Dave was amused by it.

References

Garibaldi, S. (2008). Someone more than governors need to know about trigonometry? *Mathematics Magazine* **81**, pp. 191–200.

Gasarch, W. and Kruskal, C. (1999). When can one load a set of dice so that the sum is uniformly distributed? *Mathematics Magazine* **72**, pp. 133–138.

Gasarch, W., Washington, L., and Zbarsky, S. (2018). The coefficient choosing game, *Journal of Combinatorial Number Theory* ArXiv preprint arXiv:1707.04793.pdf.

Lamport, L., Shostak, R., and Pease, M. (1982). The Byzantine general's problem, in *ACM Transactions on Programming Languages and Systems*, pp. 382–401.

Chapter 6

Gathering for Gardner
Recreational vs. Serious Mathematics

Prior Knowledge Needed: None.

6.1   Point

Martin Gardner wrote a column for Scientific American on *Mathematical Recreations* from 1956 until 1981. His column inspired me and others to study math. He also had an interest in magic and in scientific literacy (or perhaps illiteracy).

Every two years there is the Gathering for Gardner which, is a meeting to celebrate his legacy. They are numbered G4G1, G4G2,.... The latest was G4G13.

It's a strange and delightful mix of mathematicians, scientists, and magicians. Within each field there are amateurs, professionals, and in-betweens. There are assorted other people as well (e.g., Darling). The conference can be very inspiring, and provide ideas for projects. Does that actually happen? See the last section of this chapter for an example.

While some of the math Gardner described was clearly recreational, and some was clearly serious, it was sometimes hard to tell the difference. Sometimes serious math comes out of recreational math.

Michael Henle and Brian Hopkins [Henle and Hopkins (2013)] have a edited a collection of articles that involve "serious" math inspired by Gardner's "recreational math." Rather than try to figure out which is which, I would urge you to read the book. Or save time by reading my review of it [Henle and Hopkins (2013)].

This chapter will describe some of the talks at G4G12. All of the talks are on YouTube. Hence, if one of my descriptions intrigues you then go to YouTube and type in the title and/or author and/or Gardner[1] Alternatively some of the talks are now papers available on the web. There have been proceedings for some of the gatherings. I have written a review of six of them [Gasarch (2011)].

Rochelle Kronzek, my acquisitions editor at World Scientific Publishing, got me an invitation to the conference. Without further ado I will describe some of the talks.

6.2   The First Batch of Talks from G4G12

64=65 and Fibonacci, as Studied by Lewis Caroll, by Stuart Moshowitz.

This was about a famous Lewis Caroll puzzle [Yorgey (2018)] where he put together shapes in one way to get a rectangle of area 65, and another way to get a square of area 64. How did he do it!?

How Math Can Save Your Life, by Susan Marie Frontczak.

This was a talk about bricks and weights and then she stood on the desk and sang a song [Frontczak (2017)]. I can't describe it, so go to the bibliography reference, which includes a link to the YouTube video.

---

[1]This may not always work. For some of the talks below I did not know the author and could not find it on YouTube.

Twelve Ways to Trisect an Angle, by David Richeson.

This was not a talk about cranks who thought they had trisected an angle with straightedge and compass. It was about people who used a straightedge, compass, and *just one more thing* to trisect an angle. I asked David later if the people who trisected the angle before it was shown impossible had a research plan to remove the *one more thing* and get the real trisection. He said no — people pretty much knew it was impossible even before the proof.

The Sleeping Beauty Paradox Resolved, by Pradeep Mutalik.

Clyde and I tried to explain this paradox but got confused. The Wikipedia reference is pretty good. We encourage you to go read it. Pradeep has a way of resolving it [Mutalik (2016)].

Larger Golomb Rulers, by Tomas Rokicki.

A Golomb Ruler is a ruler with marks at integer positions so that all of the distances between the marks are different. The number of marks is *the order of the ruler*. The distance between the first mark (traditionally 0) and the last mark is *the length of the ruler*.

Example: A ruler with marks at $0, 1, 2, 4, 8$ is a Golomb ruler of order 5 and length 8.

Example: A ruler with marks at $0, 1, 2, 2^2, \ldots, 2^{n-1}$ is a Golomb ruler of order n and length 2^{n-1}.

Goal: Given m, find the shortest-length Golomb Ruler of order m.

Example: For $m = 5$ there are two Golomb rulers of length 11:

$\{0, 1, 4, 9, 11\}$

$\{0, 2, 7, 8, 11\}$

There are none with shorter length.

Early on there was some math used to get results, but now it is more computer searches.

Chemical Pi, by John Conway.

There are people who memorize the first n digits of π for some large n. John does something else. He has memorized the digits of π and the chemical elements in the following way:

HYDROGEN 3.141592653 HELIUM 5897932394 (next 10 digits of π) LITHIUM, etc.

That is, he memorized the digits of π by groups of 10 and separated them by the chemical elements in the order they are on the Periodic table. He claims this makes it easier to answer questions like: What is the 87th digit of pi. He also claims it gives a natural stopping point for how many digits of pi you need to memorize (need?).

While pondering this, I went to the web to find mnemonic devices for both π and the Periodic Table. There are some absurdly long ones that would be hard to memorize. Here are two short ones:

- For π look at the length of the words:

  Now I need a drink, alcoholic in nature, after the tough chapters involving quantum mechanics.

- For the Periodic table look at the first letters of these words.

  How he lies because boys can not open flowers.

  This might not work so well if you don't know whether "because" is for *Beryllium* or *Boron* or *Barium* or *Bismuth* or *Bohrium* or *Berkelium*. It's *Beryllim*, and *boys* is *Boron*. The because-Beryllim connection is reasonable since Beryllim is the only element that begins BE. The boys-Boron connection is less clear since it really could be Bohrium.

When I blogged about this some people emailed me that they thought someone as brilliant as Conway shouldn't work on this trivial stuff. I think they miss the point! The talk and the approach are fun!

6.3 Some More On Gathering for Gardner

Playing Penney's Game with Roulette, by Robert Vallin.

Penney's game is the following: Let k be fixed. Alice picks a sequence from $\{H, T\}^k$. Bob sees that sequence and then he picks a sequence from $\{H, T\}^k$.

They flip a coin until either Alice's or Bob's sequences shows up. What is Bob's best strategy? What is Alice's? Below is Alice's choice followed by Bob's best response.

Alice's choice	Bob's Choice	Prob Bob Wins
HHH	THH	$\frac{7}{8} = 0.875$
HHT	THH	$\frac{3}{4} = 0.750$
HTH	HHT	$\frac{2}{3} = 0.666$
HTT	HHT	$\frac{2}{3} = 0.666$
THH	TTH	$\frac{2}{3} = 0.666$
THT	TTH	$\frac{2}{3} = 0.666$
TTH	HTT	$\frac{3}{4} = 0.750$
TTT	HTT	$\frac{7}{8} = 0.875$

This talk looked at what happens if instead of coins you use a roulette wheel, so there are *many* more possibilities.

New Polyhedral Dice, by Robert Fathauer, Henry Segerman, and Robert Bosch.

Most people (or maybe it's just me) think that the only way to make fair dice is with platonic solids. So only 4-sided, 6-sided, 8-sided, 12-sided, and 20-sided dice can be made. Not so. These authors talked about dice that do not have to be platonic solids. They didn't just prove theorems. They actually had dice, which they sold! The most number-of-sides was 120. (My darling bought some of their dice.)

Secret Messages in Juggling and Card Shuffling, by Erik Demaine.

I met four other theoretical/math people at the gathering. Lance Fortnow, Dana Randall, Peter Winkler, and Erik Demaine. Lance and Dana are from Georgia Tech and hence were local; I doubt they would come otherwise. Peter Winkler and Erik Demaine are really into puzzles and games, so I was not surprised they were there. I note that Erik is so well rounded that calling him a *theoretical/math person* doesn't seem quite right.

And now back to the talk. Erik and his father Martin have made up many fonts [Demaine and Demaine (2015b)] including a Juggling Font [Demaine and Demaine (2015a)], which is awesome.

Fibonacci Lemonade, by Andrea Johanna Hawksley.

The authors made lemonade by putting layers of lemon and sugar in Fibonacci number increments. Layer n has $F(n-1)$ parts lemon and $F(n-2)$ parts sugar. They also use food coloring so you can see the layers!

Penis Covers and Puzzles: Brain Injuries and Brain Health, by Gini Wingard-Phillips.

Gini recounted having various brain injuries and how working on mathematical puzzles, of the type Martin Gardner popularized has helped her recover! As for the title — people with brain injuries sometimes have a hard time finding the words for things so they use other words. In this case she wanted her husband to buy some *condoms*, but couldn't think of the word, so she said *penis covers* instead.

Loop — Pool on an Ellipse, by Alex Bellos.

Similar in my mind to the Polyhedral dice talk (you'll see why). We all know that an elliptical pool table with a hole at one of the focii has the following property: if the ball is placed at the other focii and hit hard enough it, in any direction, it *will* go into the other hole. But Alex actually *makes* and *sells* these pool tables. When he talked about this, in 2016, he was selling them for $20,000. When I checked the website in 2018 the price was (wisely?) missing. Someone made them in 1962 but few people bought them. I wonder how many are going to buy his.

Alex had problems with friction since this only works on a friction-less surface. Hence hitting the ball into the hole actually requires some skill. I do not know if this is a bug or a feature.

The similarity to dice is that I (and you?) are used to thinking about dice and ellipses abstractly, not as objects people actually build.

The pamplets of Lewis Caroll: Games, Puzzles, and Related Pieces, by Christopher Morgan.

This was mostly about puzzles that are by now familiar. I was struck by one I had not seen: an *aloof word* is a word where if you change any one letter to anything else then its no longer a word. I think aloof is such a word.

PhiTop and the Superegg, by Kenneth Brecher.

Both the PhiTop and the Superegg are three dimensional objects that look like they should wobble and fall down, but they don't. The Superegg is almost an ellipsoid. Its equation in the plane is

$$(x/a)^{2.5} + (y/a)^{2.5} + (z/b)^{2.5} \leq 1$$

Piet Hein invented it. Among many other accomplishments he also wrote several books of *Grooks* which are witty sayings that rhyme. Here is my favorite:

Problems Worthy of Attack
Prove Their Worth by Hitting Back

Clyde's favorite Grook involves indecision:

Whenever you're forced to make up your mind
And you're troubled by not having any,
The best way to solve the dilemma, you'll find
Is simply by spinning a penny.

No — not so that chance shall decide the affair
While you're passively standing there moping,
But the moment the penny is up in the air
You will suddenly know what you're hoping.

Black Hole Numbers (Author unknown, at least to me),

If you have a rule that takes numbers to numbers, are there numbers that ALL numbers eventually go to? If so, they are *black hole numbers* for that rule.

Example

$f(n)$ is the number of letters in the name of n.
$f(20) = 6$ since *twenty* has 6 letters.
$f(6) = 3$ since *six* has 3 letters.
$f(3) = 5$ since *three* has 5 letters.
$f(5) = 4$ since *five* has 4 letters
$f(4) = 4$ since *four* has 4 letters.
Note that the sequence stops here.

It turns out that ANY number eventually goes to 4. Can you prove this? I hope so, since I can't.

Another Example

$f(n)$ is the sum of the digits of n's divisors
$f(12) = 19$ since 12 has divisors 1,2,3,4,6,12 and $1 + 2 + 3 + 4 + 6 + 1 + 2 = 19$
$f(19) = 11$ since 19 has divisors 1,19 and $1 + 1 + 9 = 11$
$f(11) = 3$ since 11 has divisors 1,11 and $1 + 1 + 1 = 3$
$f(3) = 4$ since 3 has divisors 1,3 and $1 + 3 = 4$
$f(4) = 7$ since 4 has divisors 1,2,4 and $1 + 2 + 4 = 7$
$f(7) = 8$ since 7 has divisors 1,7 and $1 + 7 = 8$

$f(8) = 15$ since 8 has divisors 1,2,4,8 and $1 + 2 + 4 + 8 = 15$

$f(15) = 15$ since 15 has divisors 1,3,5,15 and $1+3+5+1+5 = 15$

It turns out that ANY number eventually goes to 15. Can you prove this? I hope so, since I can't. This example is more natural than the prior one since the prior one dealt with the English names of numbers, which is not mathematical property.

Boomerang Fractions (Author unknown, at least to me).

Let $x_1 = \frac{4}{3}$. Given x_i you get to decide if

- $x_{i+1} = x_i + \frac{1}{3}$, or
- $x_{i+1} = \frac{1}{x_i}$

your goal is to get back to $\frac{4}{3}$ as soon as possible.

To generalize you can start with any value x_1 (so $x_1 = \frac{4}{3}$ above) and, at each iteration, add a constant f (so $f = \frac{1}{3}$ above) or invert.

Almost nothing is known about this problem. So go work on it!

Liar/Truth Teller Patterns on a Square Plane, by Kotani Yoshiyuki.

You have a 4×4 grid. Every grid point has a person. They all say *I have exactly one liar adjacent (left, right, up, or down) to me.* The liars always lie, the truth tellers always tell the truth. How many ways can this happen? This can be massively generalized.

Speed Solving Rubik's Cube, by Van Grol and Rik.

The authors built a robot that can solve Rubik's cube in 0.9 seconds. By contrast the human record as of 2018 is a sluggish 4.22 seconds.

6.4   I Was Inspired

I said in the introduction that the gathering *could* be inspiring. But is it? In 2016, at G4G12, I saw a pamphlet which contained *The Muffin Problem*. In 2018, at G4G13, I gave a talk on *The Muffin Problem*. I summarize the slides:

In 2016 I saw The Muffin Problem in a Pamphlet at G4G12:

m muffins and s students. Divide and distribute maximizing smallest piece

Yada Yada Yada [for five minutes]

I now have:

(1) Eight co-authors.
(2) A 200 page paper (google *Gasarch Muffin*).
(3) A book contract.
(4) For $1 \leq s \leq 60$ and $1 \leq m \leq 70$ have determined answer for m muffins and s students. Example: 43 muffins, 33 students,

   • There is a procedure with smallest piece $\frac{91}{264}$.
   • There is no procedure with smallest piece $> \frac{91}{264}$.

(5) Many theorems, techniques, and conjectures.

Is the muffin problem serious or recreational mathematics?

References

Demaine, E. and Demaine, M. (2015a). Juggling font, *Mathematical and puzzle fonts and typefaces* http://erikdemaine.org/fonts/juggling/.

Demaine, E. and Demaine, M. (2015b). Mathematical and puzzle fonts and typefaces, http://erikdemaine.org/fonts/.

Frontczak, S. (2017). How math can save your life, https://www.youtube.com/watch?v=G53JoP3HR00.

Gasarch, W. (2011). Review or proceedings from Gathering for Gardner, *SIGACT News* **42**, 1, `https://www.cs.umd.edu/users/gasarch/BLOGPAPERS/gardnergath.pdf`.

Henle, M. and Hopkins, B. (eds.) (2013). *Martin Gardner in the 21st century* (MAA).

Mutalik, P. (2016). Solution: Sleeping beauty's dilemma, *Quanta* `https://www.quantamagazine.org/solution-sleeping-beautys-dilemma-20160129`.

Yorgey, B. (2018). The area paradox, *The math less traveled* `https://mathlesstraveled.com/2011/05/02/an-area-paradox/`.

Chapter 7

Timing is Everything! Every Planar Graph is 4.5-Colorable

Prior Knowledge Needed: Chromatic number of a graph.

7.1 Point

Lets say you prove a theorem. Yeah! How much others care may depend on the timing. We give an example of this related to fractional graph colorings.

7.2 Every Planar Graph is 4.5-Colorable — A Personal History

In 1990 I went to a talk by Jim Propp on the fractional chromatic number of a graph. We defer the definition of fractional coloring of a graph to the next section. A good reference for fractional graph theory is a book by Ed Scheinerman and Dan Ullman [Scheinerman and Ullman (1996)].

The big open problem about fractional graph colorings, and perhaps the motivation for the field, was this:

- The following is known:
  Every planar graph is 5-colorable.
  This was proven in 1890 by Heawood drawing heavily on earlier work by Kempe. The proof is in most texts on graph theory (and is human-readable).

- The following is known:
  Every planar graph is 4-colorable.
  The proof, in 1976, by Appel and Haken [Appel and Haken (1977a,b,c, 1989)] used computers to check many cases. In 1996, a simpler and shorter proof was found by Robertson, Sanders, Seymour, and Thomas [Robertson *et al.* (1997)]; however, their proof, which still relied on a computer checking of cases, is not human-readable.
- Open: Is there a human-readable proof that
  Every planar graph is 4-colorable.
- Obtaining a human-readable proof of the four-color theorem seemed hard. What if we try for an easier goal. The five-color theorem is already known to have a human readable proof. Hence people defined a notion of *fractional chromatic number*. With that notion, the following open problem was stated:

Is there a number c, $4 < c < 5$, such that there is a human-readable proof of the theorem:
Every planar graph is c-colorable.

I had not thought about this problem for a long time. Then, in December 2015, I noticed that, at the *Discrete Math Seminar in Virginia Commonwealth University*, there was a talk on *Fractional Chromatic Number by Dan Cranston*. I emailed Dan asking about it:

Bill: *Fractional colorings! Has there been progress on finding a reasonable proof that every planar graph is c colorable for some number less than 5?*

Dan: *Yes. By Landon Rabern and me. We have shown that every planar graph is 4.5-colorable [Cranston and Rabern (2014)].*

Bill: *That can't be true. If that were true I would know it (see Chapter 24).*

Dan: *Is that a rigorous proof technique?*

Bill: *Hmmm, I guess not. Okay, GREAT! I'm surprised that I haven't heard the result. So why is it not well known?*

Dan: *The paper is on arXiv but not published yet. Also, while you and I think it's a great result, since its already known that all planar graphs are 4-colorable, most people are not interested.*

Bill: *Too bad you didn't prove it in 1975.*

Dan: *I was in kindergarten.*

Bill: *Were you good at math?*

Dan: *My finger paintings were fractional colorings of the plane and I never used more than 4.5 colors.*

Some thoughts:

(1) I would have thought that they would first get something like 4.997 and then whittle it down. No. It went from 5 right to 4.5. Reducing it any further looks hard and they proved that it cannot be a tweak of their proof.

(2) The paper is readable. It's very clever and doesn't really use anything not known in 1975. But the paradigm of fractional colorings wasn't known, and, of course, the right combination of ideas would have been hard to find. In particular, the method of discharging is better understood now, partly because of the proof of the Four Color Theorem.

(3) When I told this to Clyde, he wanted to know if there was a rigorous definition of *human-readable* since the open question asked for a human-readable proof. I doubt that there is,

or can be, but Dan's paper clearly qualifies. We can define a paper to be *human-readable* if two humans have actually read it and understood it. Or perhaps you can parameterize: A paper is *n-human-readable* if n humans have read and understood it.

7.3   The Definition of Fractional Coloring

We first do an example. Take C_5 the cycle on 5 vertices. This requires 3 colors; however, RED is used twice, BLUE is used twice, and GREEN is used once. Somehow this seems close to being 2-colorable. Instead of having 3 colors and assigning 1 to each vertex, we consider the following: there are 5 colors and we assign 2 to each vertex, insisting that two adjacent vertices have disjoint sets of colors. We can do this!:

Vertex 1 is assigned $\{1, 2\}$.
Vertex 2 is assigned $\{3, 4\}$.
Vertex 3 is assigned $\{1, 5\}$.
Vertex 4 is assigned $\{2, 3\}$.
Vertex 5 is assigned $\{4, 5\}$.

More generally, a graph is $(a : b)$-colorable if you can assign to every vertex a b-sized subset of $\{1, \ldots, a\}$ such that two adjacent vertices are assigned disjoint sets. One is tempted, then, to define a graph to be a/b-colorable if it is $(a : b)$-colorable. But we would need that G is $(a : b)$-colorable iff G is $(ka : kb)$-colorable. The implication is obvious; however, the converse is not. The following definition has been used:

Definition 7.1. Let G be a graph and $b \in \mathbb{N}$. $\chi_b(G)$ is the least a such that G is $(a : b)$-colorable. The *fractional chromatic number of* G is $\lim_{b \to \infty} \frac{\chi_b(G)}{b}$.

Is this the right definition? This is not a mathematics question; however, one way to test if a definition is the right one is to see if it is equivalent to other definitions. We define another notion of fractional chromatic number and then state (but not prove) that they are equivalent.

Consider the ordinary chromatic number of a graph $G = (V, E)$. We want to find the minimal number of independent sets that cover the entire graph. We phrase this as a 0-1 programming problem. For every independent set I of G we have a variable x_I. We intend to set x_I to 1 if it's included in our set of independent sets, and 0 otherwise.

Definition 7.2. The *chromatic number of G* is the minimal value of

$$\sum_{I \text{ an ind set}} x_I$$

relative to the following constraints:

- For every vertex v,

$$\sum_{I \text{ such that } v \in I} x_I \geq 1.$$

(This guarantees that every vertex is in some I.)
- $x_i \in \{0, 1\}$. Note that x_i is $\{0, 1\}$-valued.

The above formulation of chromatic number inspires the following definition where we relax the constraint that $x_i \in \{0, 1\}$.

Definition 7.3. The *fractional chromatic number of G* is the minimal value of

$$\sum_{I \text{ an ind set}} x_I$$

relative to the following constraints:

- For every vertex v,

$$\sum_{I \text{ such that } v \in I} x_I \geq 1.$$

- $x_i \in [0, 1]$. Note that x_i is now real-valued.

Theorem 7.1. *The two definitions of fractional chromatic number are equivalent.*

I believe the definitions given are the right ones because they are both reasonable and they are equivalent. Clyde disagrees because the fractional chromatic number might not be within 1 of the chromatic number. For example, Dan Cranston and Landen Rabern [Cranston and Rabern (2015)] show the following:

Theorem 7.2. *The fractional chromatic number of the plane is between 3.5556 and 4.3599.*

(A good reference for material on chromatic number of the plane is a book by Alexander Soifer [Soifer (2009)].) It is known the chromatic number of the plane is between 5 and 7. However, most people in the know think the chromatic number of the plane is 6 or 7. Hence this is likely a case where the fractional chromatic number is much less than the chromatic number.

References

Appel, K. and Haken, W. (1977a). Every planar map is four colorable I: Discharging, *Illinois Journal of Mathematics* **21**, pp. 429–490.

Appel, K. and Haken, W. (1977b). Every planar map is four colorable II: Reducibility, *Illinois Journal of Mathematics* **21**, pp. 491–567.

Appel, K. and Haken, W. (1977c). Solution of the four color map problem, *Scientific American* **237**, pp. 108–121.

Appel, K. and Haken, W. (1989). *Every planar map is four colorable* (AMS).

Cranston, D. and Rabern, L. (2014). Planar graphs are 9/2-colorable and have independence ratio at least 1/3, Paper: http://arxiv.org/abs/1410.7233, talk: http://www.people.vcu.edu/~dcranston/slides/ planar-fractional-SFU.pdf.

Cranston, D. and Rabern, L. (2015). The fractional chromatic number of the plane, Paper: http://arxiv.org/ abs/1501.01647, talk: http://www.people.vcu.edu/ ~dcranston/slides/plane-fractional.pdf.

Robertson, N., Sanders, D., Seymour, P., and Thomas, R. (1997). The four color theorem, *Journal of Combinatorial Theory, Series B* **70**, pp. 2–44.

Scheinerman, E. and Ullman, D. (1996). *Fractional graph theory* (Dover), http://www.ams.jhu.edu/ers/books/fractional-graph-theory-a-rational-approach-to-the-theory-of-graphs/.

Soifer, A. (2009). *The mathematical coloring book: mathematics of coloring and the colorful life of its creators* (Springer-Verlag, New York, Heidelberg, Berlin).

Chapter 8

Ramsey Theory and History: An Example of Interdisciplinary Research

Prior Knowledge Needed: None.

8.1 Point

When people in pure math tell me that *you never know when something will apply to the real world*, I am skeptical. But recently I found an application of Ramsey Theory (yes, Ramsey Theory!) to History (yes, History!). After eating a slice of humble pie[1], I wrote this survey of the history of the application of Ramsey Theory to history. It is self contained.

8.2 Introduction

The application of mathematics to the natural sciences, and the natural sciences as a source of interesting problems in mathematics, is a well known phenomenon. The famous physicist, Eugene Wigner [Wigner (1960)], thought the application of math to the natural sciences was *unreasonably effective*. What about the application of mathematics to the social sciences? Economist Herbert Scarf [Scarf (1983)] applied the Brouwer fixed point

---

[1]The only people whom I've ever heard use the expression "eat humble pie" or "eat crow" are characters on TV.

theorem to economics. This is one of many examples of applying (arguably) pure mathematics to economics. Samuel Huntington's use of mathematics in political science caused much controversy [Simon (1988); Koblitz (1988); Simon and Koblitz (1988)] when he was nominated to the National Academy of Sciences (he was rejected).

What about sociology or history? The application of statistics to these fields is well known. Dorwin Cartwright, Frank Harary [Cartwright and Harary (1977)], and other mathematicians, appear to have used graph theory to model social relationships; however, on closer inspection they just used the *language of graph theory*. While this is all well and good, the question arises: has *pure mathematics* ever been used in a serious way on a problem of sociology or history? Chemist and novelist C. P. Snow [Snow (1959)] wrote a famous article, *The Two Cultures*, that speculates that there is a cultural divide between the sciences and the humanities which may make such collaborations difficult. This points to the lack of interaction being a sociological problem in itself; however, we are not going to go there. While there may really be such connections between sociology and pure mathematics, they will be hard to find.

There was *an* application of Ramsey theory to sociology in the 1950s. In Jacob Fox's *Lecture Notes in Combinatorics* [Fox (2009)] he tells the following true story:

> In the 1950s, a Hungarian sociologist Sandor Szalai studied friendship relationships between children. He observed that in any group of around 20 children he was able to find four children who were mutual friends, or four children such that no two of them were friends. Before drawing any sociological conclusions, Szalai consulted three eminent mathematicians in Hungary at that time: Paul Erdős, Paul Turán, and Vera Sós. A brief discussion revealed that indeed this is a mathematical

phenomenon rather than a sociological one. For any symmetric relation R on at least 18 elements, there is a subset S of 4 elements such that R contains either all pairs in S or none of them. This fact is a special case of Ramsey's theorem, proved in 1930, the foundation of *Ramsey Theory* which developed later into a rich area of combinatorics.

This could be called an anti-application since, in the end, there was no interesting sociological phenomena. And, while this story is amusing (and surprisingly true), sociology did not become a source of questions for mathematics.

Recently there *has* been a case where Ramsey Theory greatly simplified a topic in history. Conversely, history has been the source of interesting new problems in Ramsey Theory. Surprisingly, history has helped solve an open problem in Ramsey Theory. This paper is an exposition of these events.

8.3   Pre-Christian History of England

Sir Woodsor Kneading is a scholar of pre-Christian English history. He is particularly interested in when wars (technically skirmishes) broke out. He noticed the following:

(1) In 577 BC there were five lords in what is now West Essex, all with armies. Some of the pairs were allied and some were enemies, yet there was no war. Then a sixth Lord settled into that region and within a year war broke out.

(2) In 552 BC there were five lords in what is now South Wales, all with armies. Similarly, once a sixth Lord settled, there was war.

(3) He noticed this happening 42 times total.

He then looked for this pattern and discovered the following:

*Between the years 600 BC and 400 BC, whenever
there were six lords in proximity war broke out, with one
exception. That one exception was truly exceptional —
that was when all six lords had an alliance with each
other. The question arises: Why do six lords almost
always mean war?*

*I believe I have an answer. I noticed that either
(1) three, four, or five of them formed an alliance and,
thinking themselves quite powerful, merged armies and
attacked the other lords, or (2) there were three or more
of them who were pairwise enemies, and in that case war
broke out among these factions, or (3) (and this is rare)
all six formed an alliance and there was peace. Note that
the wars in cases (1) and (2) were very different from
each other, but they were still wars.*

Kneading goes on:

*I conjectured the following: whenever there are six
lords, not all in alliance, there must either be (1) three,
four, or five who are all allied with each other, or (2)
three, four, five, or six who are pairwise enemies. I hired
a computer science undergraduate student H. K. Donnut
to look into this. After a month he proved my conjecture by a clever computer search. We published a joint
paper [Kneading and Donnut (2011b)].*

After the year 400 BC there were cases of six lords in a region and no war. Why was this? Kneading speculated that
around that time weapons became more high tech so wars became more costly. This speculation was verified by Moss Chill
Beaches [Beaches (2013)].

Kneading then noticed that, between the years of 400 BC
and 200 BC, whenever there were at least 18 lords in the same
region there was a war.

Between the years of 400 BC and 200 BC, whenever there were at least 18 lords in proximity either (1) between four and seventeen of them formed an alliance and, thinking themselves quite powerful, merged armies and attacked the other lords, or (2) there were four or more of them who were pairwise enemies, and in that case war broke out among these factions. If ever there was a time when all 18 had alliances between them then peace could have broken out but, alas, this never happened.

Once again Kneading hired Donnut to look into this. After two months, they wrote a paper [Kneading and Donnut (2011a)] whose title gives away the main theorem: *18 lords means war!*.

8.4   Ramsey Theory

We state Ramsey's Theorem for graphs in a way that it can be applied to history easily.

(1) K_n is the complete graph on n vertices. We will think of the vertices as being lords.

(2) We will be 2-coloring the edges of K_n. If two lords are enemies then we draw a RED line between them (RED for blood). If two lords have an alliance then we draw a BLUE line between them (BLUE for friendship).

(3) Let COL be a 2-coloring of the edges of K_n. A subset of vertices U of size k is called a *k-alliance set* if all of the lords in U have an alliance with each other. A subset of vertices U of size k is called a *k-enemy set* if all of the lords in U are pairwise enemies.

Note 8.1. Kneading and Donnut called these colored graphs *Alliance-Enemy Descriptors* or AEDs. Graph theorists would simply call these *a 2-coloring of the edges of K_n*. In addition, graph theorists would call a k-enemies set *a red clique on k vertices* and a k-alliance set *a blue clique on k vertices.*

It turns out that Woodsor and Donnut [Kneading and Donnut (2011b,a)] rediscovered the following theorem:

Theorem 8.1.

(1) For any 2-coloring of K_6 there is either a 3-enemies set or a 3-alliance set. (The only case where 6 lords did not mean war occurred was when there was a 6-alliance set, which was rare in the time period considered.)

(2) For any 2-coloring of K_{18} there is either a 4-enemies set or a 4-alliance set. (The only case where 18 lords would not mean war would be if there was an 18-alliance, but this never happened in the time period considered.)

Had Kneading known these theorems he would have saved time. If Donnut had known these theorems he would not have gotten paid for three months of work, nor would he have been a co-author on the two papers.

8.5   Kneading's Book

Kneading and Donnut kept on working on history viewed in terms of AEDs. They ended up rediscovering several known small Ramsey Numbers including bipartite ones (useful if the lords live on different sides of a river). Kneading wrote an entire book [Kneading (2013)] on this topic without knowing any Ramsey theory. (Donnut declined to be a co-author as he thought of himself as merely being a coder whereas Kneading had done the intellectual heavy-lifting.)

This book was a breakthrough in the study of war and peace. It was largely because of this book that Kneading was elected a foreign member of the (American) National Academy of Sciences [Grant (2015)].

We present the general Ramsey Theorem for 2-colorings of graphs and then a passage from Kneading's book. We believe the passage can be construed as stating Ramsey's Theorem. We state the theorem the way a mathematician would.

Theorem 8.2. *For all k there exists n such that for all 2-colorings of the edges of K_n there is a monochromatic k clique.*

The Relevant Passage from Kneading's Book:

> *As wars get more and more costly, the number of lords (or in the modern world, countries) that need to either align or be pairwise enemies to guarantee war, will increase. Is there a limit? I doubt it. If, for example, a 17-enemies set or 17-alliance set is needed to start a war then I expect there is some number of lords, perhaps quite large, so that if that number of lords were in close proximity, a war would break out.*

8.6 Simplification Due to Ramsey's Theorem

Alma Grand-Rho was visiting Cambridge, where Kneading teaches, to give a talk on her ideas for a simplification of D.H.J. Polymath's proof of the Density Hales-Jewitt Theorem [Polymath (2012)]. She lost her way and found herself in the History department. The department had one of those bookshelves where books of faculty are displayed. She noticed Kneading's book *Alliances, Enemies, and War in Pre-Christian England*, which intrigued her since she had minored in history and did a senior thesis on the English kings. While browsing the book she found the following passage:

> *In the year 100 AD, in West Wales, there were 48 lords in proximity but no war broke out. In this time period whenever five lords formed either a 5-alliance or*

a 5-enemies set there was a war. I realized that either (1) there was a 5-alliance or 5-enemies set yet no war broke out, which would be interesting and perhaps cause me to change my timeline as to when 48 was enough, or (2) there was no 5-alliance or 5-enemies set, which would be less interesting. I had Donnut enter the data and run his program to see if there were any 5-alliance or 5-enemies sets. There were not. Oh well.

She then thought: *OMG!*[2] *He just proved* $R(5) = 49$. What does this mean? It was known (by mathematicians) that every AED of size 49 has either a 5-alliance or a 5-enemies set, but it was not known for 48. Since the passage above describes a graph on 48 vertices with no 5-alliance or 5-enemies set we now know that 49 is optimal. Later on, Kneading and Grand-Rho met and got out a paper [Grand and Kneading (2014)] that essentially summarized his book succinctly using Ramsey Theory. Kneading says, without embarrassment, *My paper with Alma says cleanly in 30 pages what I said clumsily in 300 pages.*

Their paper was a lucid explanation of concepts in both Ramsey Theory and history. For this paper they were awarded both the Steele Prize for Mathematical Exposition [Writset (2015)] and the Pulitzer prize for History.

8.7   Open Problems

Kneading's model of alliance and enemies might be too simple to grapple with today's countries and other groups. At the conference celebrating noted Ramsey theorist Tee A. Cornet's 100th birthday, Grand-Rho and Kneading [Grand and Kneading (2015)] presented the following refinements.

(1) Some pairs of lords are enemies and some are allies. This

---

[2]OMG is text-speak for Oh My God.

seems too black-and-white. There could be Fifty Shades of Grey [James (2011)]. This may lead to applications of, and open problems in, multicolor Ramsey Theory.

(2) There may be alliances between three, four, or even more lords. However, there is no such thing as a set of three lords that are enemies except to say it is pairwise. This may lead to a new type of Ramsey Theory where you color (say) edges RED or BLUE, but hyperedges are either colored BLUE or not colored at all.

(3) You can combine the two above to get a new type of multicolor, hypergraph Ramsey Theory.

(4) Kneading assumed that alliances and enemies are symmetric. There can be cases where A likes B but B does not like A. Hence one can look at asymmetric Ramsey Theory. Again, one may combine this with the above to get asymmetric, multicolored, hypergraph Ramsey Theory.

In summary, the story I tell above is strong evidence of the value of interdisciplinary research. Ramsey Theory simplified some work in history, and history is now the impetus for new work in Ramsey Theory. And, of course, most surprisingly, historians solved an open problem in mathematics.

8.8   Acknowledgements

I would like to thank Moss Chill Beaches, H. K. Donnut, Alma Grand-Rho, and Sir Woodsor Kneading for helpful discussions. I am especially thankful to Kneading for his honesty about how Ramsey Theory made his work much cleaner. He is a model of academic honesty and grace.

I would like to thank Andy Parrish for pointing me to Sandor Szalai's work which I consider a precursor to the work surveyed in this chapter. I would like to thank Stephen Fenner, Clyde Kruskal, and Andy Parrish for proofreading and helpful discussion.

References

Beaches, M. C. (2013). Why six lords stopped meaning war, *Journal of Pre-Christian English History* **63**, pp. 15–48.

Bergelson, V. and Leibman, A. (1996). Polynomial extensions of van der Waerden's and Szemerédi's theorems, *Journal of the American Mathematical Society* **9**, pp. 725–753, `http://www.math.ohio-state.edu/~vitaly/` or `http://www.cs.umd.edu/~gasarch/vdw/vdw.html`.

Cartwright, D. and Harary, F. (1977). A graph theoretic approach to the investigation of system-environment relationships, *Journal of Mathematical Sociology* **5**, pp. 87–111.

Fox, J. (2009). Lecture notes in combinatorics, `http://math.mit.edu/~fox/MAT307-lecture01.pdf`.

Grand, A. R. and Kneading, S. W. (2014). Applying Ramsey theory to History and History to Ramsey theory, *Journal of Mathematical History and Historical Mathematics* **5**, pp. 6–36.

Grand, A. R. and Kneading, S. W. (2015). Old and new problems and results in Ramsey-theoretic history, in *Easy and Hard Ramsey Theory (In honor of Tee A. Cornet's hundredth birthday)* (Publish or perish press), pp. 43–49.

Grant, A. T. (2015). Kneading elected to National Academy of Sciences, `http://www.nasonline.org/`.

James, E. L. (2011). *Fifty shades of Grey* (Vintage Books).

Kneading, S. W. (2013). *Alliances and Enemies in Pre-Christian England* (Publish or Perish Press).

Kneading, S. W. and Donnut, H. K. (2011a). Eighteen lords mean war! *Journal of Pre-Christian English History* **64**, pp. 42–69.

Kneading, S. W. and Donnut, H. K. (2011b). Six lords almost always mean war! *Journal of Pre-Christian English History* **62**, pp. 15–48.

Koblitz, N. (1988). A tale of three equations; or the emperors have no clothes, *The Mathematical Intelligencer* **10**, 1, pp. 4–10.

Polymath, D. (2012). A new proof of the density Hales-Jewett Theorem, *Annals of Mathematics* **175**, pp. 1283–1327, `http://arxiv.org/abs/0910.3926`.

Scarf, H. (1983). Fixed point theorems and economic analysis, *American Scientist* **71**.

Simon, H. (1988). Some trivial but useful mathematics, This seems to have been well circulated but never published. I cannot find a copy online; however, it is quoted in Neal Koblitz's rebuttal to it.

Simon, H. and Koblitz, N. (1988). Opinion, *The Mathematical Intelligencer* **10**, 2, pp. 10–12.

Snow, C. P. (1959). *The Two Cultures* (Cambridge University Press), republished in 2001.

Walters, M. (2000). Combinatorial proofs of the polynomial van der Waerden theorem and the polynomial Hales-Jewett theorem, *Journal of the London Mathematical Society* **61**, pp. 1–12, `http://jlms.oxfordjournals.org/cgi/reprint/61/1/1` or `http://jlms.oxfordjournals.org/` or `http://www.cs.umd.edu/~gasarch/vdw/vdw.html`.

Wigner, E. (1960). The unreasonable effectiveness of mathematics in the natural sciences, *Communications on Pure and Applied Mathematics* **13**, `http://www.dartmouth.edu/~matc/MathDrama/reading/Wigner.html`.

Writset, A. (2015). Kneading wins Steele prize for mathematical exposition, `http://www.ams.org/profession/prizes-awards/ams-prizes/steele-prize`.

Chapter 9

Ramsey Theory and History: An Example of Interdisciplinary Research — An Alternative View

Prior Knowledge Needed: Chapter 8.

9.1 Story and Point

In the Spring of 2013 I taught a graduate course titled *Ramsey Theory and its "Applications"*. Yes, the actual course name had quotes around the word *Applications*. This is because the applications were mostly to esoteric theorems in math or theoretical computer science.

All semester I said things like *there has been some progress on R(5), but I'll talk about that later.* In the middle of the semester I gave them a version of the paper in the chapter *Ramsey Theory and the History of Pre-Christian England: An Example of Interdisciplinary Research*.

I told them that I was going to do an experiment with the flipped classroom concept. This is a real educational innovation where the students learn the material on their own at home (often with videos but not in my case) and then in class they discuss the material. They would read the paper, and a week later, I would give them a quiz on it (to make sure they read it) before discussing it.

I did indeed give them a quiz, but I asked them to *not* put their name on it because it was actually a survey. Below, I have

99

the questions and a description of how they answered it. For example, for a multiple choice question. I indicate how many wrote each choice. Seventeen people took the quiz (the class had sixteen students and some auditors).

(1) *When did you realize the paper was a hoax?*

  (a) When I first took this quiz: 5 students.
  (b) When I saw that *Fifty shades of Grey* was one of the references: 1 student.
  (c) When I went to the website that was supposed to have R(5) and it revealed the hoax: 3 students. (Note — the version of the paper I gave them had such a website. Later I'll say why I removed it for this book.)
  (d) When I saw the name H. K. Donnut: 2 students.
  (e) When I read your April 1 blog entry: 1 student.
  (f) When I heard other students talk about it (though I kind of knew anyway): 4 students.
  (g) When I read the title: 1 student.

  UPSHOT: I would count answers a,b,c as being fooled. So 9 were fooled.

(2) *Was it okay to have you read a hoax paper?* (Darling thinks it is not okay.)

  (a) This assignment was awesome!: 9 students.
  (b) I liked learning to not always believe a professor: 1 student. (Later in the semester when I told the class that the Large Ramsey Theorem can't be proven in Peano Arithmetic he was skeptical, thinking that, once again, I was trying to fool them.)
  (c) The assignment was okay. (I could tell these students were not amused.) 4 students.
  (d) Okay as far as it goes, but 8 pages! C'mon, that's too much: 1 student (This is fair, but it was an easy read.)

(e) It was cool since it was close to April Fools Day (the quiz was given on April 2): 1 student.

UPSHOT: It was fair to do once.

(3) *Which of the names are real and which did I make up?* (They are in order of when they appeared in the article.)

   (a) Eugene Wigner. REAL. Real: 13, Fake: 4. Name does sound funny.

   (b) Herbert Scarf. REAL. Real: 8, Fake: 9. Real name that fooled the most people.

   (c) Samuel Harrington. REAL. Real: 16, Fake: 1. Sounds so real it has to be fake.

   (d) Dorwin Cartwright. REAL. Real: 11, Fake: 6. I thought this would fool more people.

   (e) Frank Harary. REAL. Real: 14. Fake: 3. One student thought I misspelled the real name. I didn't, but given my bad spelling, and the nature of the name, I can see why they thought so.

   (f) C. P. Snow. REAL. Real: 12. Fake: 5.

   (g) Jacob Fox. REAL. Real: 14. Fake: 3. I've met him and wrote a paper with him that has a total of six authors. If he does not exist then he managed to fool five co-authors.

   (h) Sandor Szalai. REAL. Real: 13. Fake: 4. Looks fake to me.

   (i) Paul Erdős. REAL. Real: 17. Fake: 0. He seems like a mythical figure when you know his life story.

   (j) Paul Turán. REAL. Real: 14. Fake: 3. One student thought it was a play on Turing.

   (k) Vera Sós. REAL. Real: 14. Fake: 3. I'm surprised people thought this name is real — I would have thought fake.

   (l) Sir Woodson Kneading. FAKE. Real: 7. Fake: 10. It's an anagram of Doris Kearns Goodwin, a presidential historian. She studies presidents. I am not saying that she is, herself, presidential. Then again, we've had presidents that were not presidential. I'm amazed that 7 thought it was real.

(m) H. K. Donnut. FAKE. Real: 4. Fake: 13. It's an anagram of Don Knuth, the father of algorithmic analysis. The name looks so fake it has to be real. But no.

(n) Moss Chill Beaches. FAKE. Real: 4. Fake: 13. It's an anagram of Michael Beschloss, a presidential historian. As with Doris Kearns Goodwin, I am not saying that he is presidential. My most fake-looking name.

(o) Tim Andrer Grant. FAKE. Real: 12. Fake: 5. It's an anagram of Martin Gardner. He had a math column in the magazine *Scientific American* for many years. *Tim* is a reasonable first name, *Grant* is a reasonable last name, and *Andrer* — well, middle names can be unusual.

(p) Alma Rho Grand. FAKE. Real: 3. Fake: 14. It's an anagram of Ronald Graham, a mathematician who has worked on Ramsey Theory and is a co-author of a book on that subject [Graham *et al.* (1990)]. My favorite fake name.

(q) D.H.J. Polymath. FAKE. Real 5. Fake: 12. While this name is fake, I did not invent it. Terence Tao (or should I say Tee A. Cornet) has had several polymath projects. These are projects where many people with different ideas work on a math problem together and if the problem is solved then the paper uses an alias for an author. The first problem so proposed was to find an elementary proof of the Density Hales Jewitt Theorem. They succeeded and they used, as an alias for the author of the paper, D.H.J. Polymath [Polymath (2012)]. The people who thought it was real were either kidding or read the question as a trick question. *Fake name that Dr. Gasarch made up* — this is a fake name but Dr. Gasarch didn't make it up.

(r) Ana Writset. FAKE. Real: 13. Fake: 4. It's an anagram of Ian Stewart, a popular writer of mathematics. This is the fake name that fooled the most students.

(s) Tee A. Cornet. FAKE. Real: 9. Fake: 8. It's an anagram of Terence Tao, a professor of mathematics who has solved very hard problems in Ramsey Theory using hard non-combinatorial techniques. Is Tee anyone's first name?

(t) Andy Parrish. REAL. Real: 16. Fake: 1. He worked with me on Ramsey Theory as an undergraduate and later got a PhD from UCSD in Math. His topic was Ramsey Theory.

(u) Stephen Fenner. REAL. Real: 16. Fake: 1. A professor of computer science. I hope he's real — he's in charge of SIGACT NEWS and I have an open problems column there.

(v) Clyde Kruskal. REAL. Real: 17. Fake: 0. A professor of computer science in my department. That may be why so many students knew he was real. Or maybe they think he invented Kruskal's algorithm. I hope he's real — he's my co-author on this book.

(4) *Speculate on how I came up with the false names.* Most left this blank. Some said anagrams, some mentioned anagram-programs on the web (I did use one), some said a random-name-generator; pointing out that Vera Sós and Sandor Szalai look like they were produced by a not-very-good random name generator. UPSHOT: As Blanch Nail Roam said: *You can fool some of the people all of the time, and all of the people some of the time, but you can't fool all of the people all of the time.*

A few other notes:

(1) Andy Parish, one of my proofreaders, thought it was true since he wanted it to be true.
(2) Douglas Ulrich, a student in my Ramsey Theory class, thought it was true because he could not imagine I would give a hoax paper to the class to read.
(3) Daniel Apon, another student in my Ramsey Theory class, knew the fake names were anagrams. He thought Dorwin Cartwright was fake and spent an hour trying to find what it was an anagram of.
(4) Noga Alon told me that an amateur (non-crank) mathematician he knows got very excited about this paper and wanted him to look at it. Noga knew it was a hoax for two reasons: (1) if $R(5)$ was known, then he would know it (an example of a proof technique discussed in Chapter 24). (2) he didn't recognize any of the names.
(5) My high school student Jessica Shi thought it odd that Professor Alma Grand-Rho would say OMG. Rather than thinking it was a hoax she thought that perhaps this was a rather young professor who was used to texting. However, later Jessica had an inkling it was a hoax when she noticed that *Fifty Shades of Grey* was referenced.

UPSHOT: It can be hard to distinguish fake names from real ones. Along those lines, what do Dolly Parton's and Clint Eastwood's names have in common? We answer that later.

9.2   Later Renditions

As recounted in the first section I gave the paper to a graduate course in April 2013. I later gave it to a junior-level course in April 2014 and to a freshman course in April 2015. Because of the order of the courses there was little chance that word would leak out that it was a hoax.

The results were pretty much the same as above but with one curious difference. An earlier version of the paper had a website to look at the graph that shows $R(5) = 49$, but when you go there it says *This Paper is a Hoax! All of the names are Anagrams of real people!*. Very few of the grad students I gave the paper to in April 2013 went to that website. About half of the Freshman in April 2015 did. It is a mystery as to what changed between 2013 and 2015. In any case, since too many people looked it up, I have removed that website from the paper.

Is it easier to go to the web wherever you are in 2015 then it was in 2013? It would not seem so. However, over time it has gotten much easier. My co-blogger Lance's daughter Annie once said to him: *Really daddy? There was a time when you had to be at home or at work to log on? The dark ages!*

Now about Dolly Parton and Clint Eastwood: Those are their real names. They are also great stage names.

References

Graham, R., Rothschild, B., and Spencer, J. (1990). *Ramsey Theory* (Wiley, New York).

Polymath, D. (2012). A new proof of the density Hales-Jewett Theorem, *Annals of Mathematics* **175**, pp. 1283–1327, http://arxiv.org/abs/0910.3926.

Chapter 10

Lets Prove Something Instead of Proving That We Can't Prove Something!!!

Prior Knowledge Needed: (1) P, NP, Barriers to resolving P vs. NP, (2) A sense of history.

10.1 Rant

Complexity theorists (including me) seem more concerned with proving that we can't prove things rather than actually proving things!!!! There have been two workshops on Barriers — reasons we cannot prove things. I saw an excellent talk by Peter Bro Miltersen where he listed as an open problem that he wanted to get a barrier result. He seemed more interested in proving that nobody could prove the result then in proving it!

We are in a rut. We seem to have the urge to show a theorem is hard to prove rather than actually proving it! To be fair, many of our conjectures are hard to prove. Even so, we need to get out of this rut. I list several problems that I think can be solved with current techniques. For each one I will say why the current barrier techniques might not apply.

Given two classes $\mathcal{C}$ and $\mathcal{D}$ we want to show that they are different. This is often hard (e.g., P and NP). Over the years two techniques have been developed to show that separating classes is hard:

- Virtually all separations and collapses only use techniques from computability that relativize. That is, if you put oracles on all of the Turing machines involved, the proofs still work. Baker, Gill, and Solovay [Baker *et al.* (1975)] showed that there are oracles that force P = NP and oracles that force P $\neq$ NP. When this happens we say *the P vs. NP question relativizes.* Many other classes have been shown to relativize. Hence it is unlikely that techniques from computability theory will resolve P vs NP (though they may be part of a solution).

• Virtually all separations that are known (e.g., constant depth poly sized circuits cannot compute parity) have a property that Rudich and Razborov [Razborov and Rudich (1997)] pinned down and called *natural.* They also show that, given some reasonable assumptions, no natural proof can separate P from NP or solve other similar problems. Roughly speaking this shows that the current techniques used to prove lower bounds on circuits will not suffice to determine if $\mathcal{C}$ and $\mathcal{D}$ are equal.

1) Prove that NP $\neq$ DTIME($2^{O(n)}$). Gasarch [Gasarch (1987)] showed that there are oracles for proper containment either way, and for incompatibility. No oracles are known that make NP = DTIME($2^{O(n)}$). Also note that NP is related to polynomials which have nice closure properties (e.g., if $p(n)$ is a polynomial then so is $p(n)^2$) where as DTIME($2^{O(n)}$) is related to linear functions which are not as robust. That might be useful. DO NOT TRY TO FIND AN ORACLE TO MAKE THEM EQUAL!!! THAT IS THE MENTALITY I WANT TO BREAK US OUT OF!!!

2) How do NL (nondeterministic log space) and P compare? Ladner and Lynch [Ladner and Lynch (1975)] showed that there are oracles to make either properly contained in the other.

However, relativized[1] space has several definitions, none of which seem satisfactory. Buss [Buss (1987)] and Ruzzo *et al.* [Ruzzo *et al.* (1982)] have worked on this. DO NOT TRY TO FIND THE RIGHT DEFINITION OF RELATIVIZED SPACE!!! JUST PROVE A CONTAINMENT EITHER WAY!!! IT DOESN'T EVEN HAVE TO BE PROPER!!! OR PROVE THEY ARE THE SAME (unlikely).

Determining how NL and P compare might be hard, partially because we do not have an intuition ahead of time of how they relate. So instead, perhaps try to prove that NL is contained in PSPACE, which is surely true. Proper containment may be too much to ask for with today's techniques but that's no reason not to try! But again, let's PROVE IT rather than REDEFINE ORACLES to prove that the problem is hard.

3) P vs NP is clearly a hard problem. Let's look at nondeterminism in a different light — for rather powerful classes and rather weak ones.

(1) **Powerful classes:** Let PR be the set of SETS that are Primitive Recursive, and NPR[2] be the set of all SETS that are Nondeterministic Primitive Recursive. Is PR = NPR? DO NOT TRY TO DEFINE PRIM REC WITH ORACLES TO OBTAIN A BARRIERS RESULT!!!! JUST SOLVE THE DAMN PROBLEM!!!

(2) **Weak classes:** Deterministic Finite Automata (DFA) and Nondeterministic Finite Automata (NFA). Do they recognize the same set of languages or not? DO NOT DEFINE DFAs WITH ORACLES!!! DO NOT TRY TO MAKE

---

[1]My spellchecker thinks that *relativized* should have a capital R. Why? These are the questions that try men's souls. Women's too.

[2]There is a rumor that NPR was named that because people who worked on it liked National Public Radio. There is a kernel of truth to the rumor but after much research I have concluded that it is not true. This may not be the final word on this important issue.

DFAs FIT INTO THE NATURAL PROOFS FRAME-
WORK!!! JUST SOLVE THE *&#ing PROBLEM!!!!

10.2   Hey Wait a Minute!

Some of my readers were puzzled and left comments like *Hey!
I thought we know that DFAs and NDFAs were of the same
strength.* Indeed we do. This was an April Fools Day post. All
of the problems above that I claimed were open were solved a
long time ago. So the statement

> *I list several problems that I think can be solved with
> current techniques*

is technically correct. Another problem: devise a logic where
you can distinguish statements that are true and those that are
just technically true.

This April Fools Day post is a reversal of a common April
Fools Day joke. Martin Gardner once had an April Fools Day
column where he claimed the four-color theorem, among other
open problems, was solved. The blog of Lipton and Regan has
used April Fools Day to claim factoring is in P (which is not
known). In short, it is common to claim that an open problem
is solved. I did the opposite: I claimed that a solved problem
is open. My readers corrected me rather than realizing it was a
joke.

We look at each question I raised:

1) It is known that NP $\neq$ DTIME($2^{O(n)}$): Assume, by way
of contradiction, that NP = DTIME($2^{O(n)}$). By the time hierarchy theorem there is a language $L \in$ DTIME($n2^{O(n)}$) $-$
DTIME($2^{O(n)}$). Let $\mathcal{A}$ be the algorithm for L that takes $n2^{O(n)}$
steps. Note for later use that on an input of length $\sqrt{n}$ $\mathcal{A}$ takes
$\sqrt{n}2^{O(\sqrt{n})} \leq O(2^n)$ steps.

Let

$$L' = \{x\$1^{|x|^2 - |x|} : x \in L\}$$

We show that $L' \in \mathrm{DTIME}(2^{O(n)})$:

(1) Input y of length n.
(2) Check if y is of the form $x\$1^{|x|^2 - |x|}$. If y is not of that form then reject. If y is of that form then let x be such that $y = x\$1^{|x|^2}$.
(3) Note that $|x| = \sqrt{|y|} = \sqrt{n}$. Run $\mathcal{A}$ on x. This takes $\sqrt{n}2^{O(\sqrt{n})} \leq O(2^n)$ steps.

Since $L' \in \mathrm{DTIME}(2^{O(n)}) = NP$, $L' \in \mathrm{NP}$. From this one can easily show $L \in \mathrm{NP} = \mathrm{DTIME}(2^{O(n)})$. This is a contradiction.

2) It is easy to show that $\mathrm{NL} \subset \mathrm{P}$ by viewing an NL-computation as the transitive closure of a directed graph. The oracles that got other relations between NL and P used a definition of oracles for space classes that is not quite right, partially because the proof of $\mathrm{NL} \subseteq \mathrm{P}$ does not relativize with this definition. Using the definition of Ruzzo-Tompa-Simon the proof that $\mathrm{NL} \subseteq \mathrm{P}$ does relativize.

What about NL and PSPACE? By Savitch's theorem PSPACE = NPSPACE. Combining this with the nondeterministic space hierarchy theorem we have

$$\mathrm{NL} \subset \mathrm{NPSPACE} = \mathrm{PSPACE}.$$

3a) There is no formal definition of nondeterministic primitive recursive. If you google *"Nondeterministic primitive recursive"* the only hits you get are to the blog post this chapter is based on. Perhaps you will now also get a hit in Google Books referring to this chapter.

Even so, we will proceed. Note that PR = DTIME(PR). Hence we define NPR = NTIME(PR). With this definition we have

$$\text{NPR} = \text{NTIME(PR)} \subseteq \text{DTIME}(2^{\text{PR}}) \subseteq \text{DTIME(PR)} = \text{PR}.$$

3b) As mentioned above, my reader pointed out it is well known that DFAs and NDFAs have the same power.

Even though this was an April Fools Day post, it brings up some serious questions:

- When are conjectures so hard that you retreat into proving they are hard to solve rather than solving them?
- What other branches of mathematics dwell on how their current techniques don't work?

The rest of this post is real. Honest!

10.3 A Barrier Identified and Overcome: The Erdős Distinct Distance Problem

The Erdős Distinct Distance Problem: Given n, what is the minimum number of distinct distances between n points in the plane? We call this value $g(n)$. Julia Gariabldi *et al.* [Gariabldi *et al.* (2010)] wrote an excellent book on the topic.

(1) Paul Erdős [Erdős (1946)] showed that $\sqrt{n} - O(1) \leq g(n) \leq O\left(\frac{n}{\sqrt{\log n}}\right)$. The first inequality would make a good high school math competition problem were it not so well known. We leave it to you. The second inequality is obtained by placing the points on a regular $\sqrt{n} \times \sqrt{n}$ grid and using some number theory. Erdős made no conjecture about $g(n)$ in this paper.

(2) Leo Moser [Moser (1952)] showed $g(n) = \Omega(n^{2/3})$. This paper claims that Erdős conjectured $(\forall \epsilon > 0)[g(n) \geq \Omega(n^{1-\epsilon})]$. It is likely that Erdős made this conjecture in a talk or in a paper listing open problems.

(3) Fan Chung [Chung (1984)] showed $g(n) = \Omega(n^{5/7})$. This paper also claims that Erdős conjectured $(\forall \epsilon > 0)[g(n) \geq \Omega(n^{1-\epsilon})]$.

(4) Fan Chung *et al.* [Chung *et al.* (1992)] showed $g(n) = \Omega\left(\frac{n^{4/5}}{\log n}\right)$. This paper, as do all subsequent papers, claims that Erdős conjectured $g(n) = \Theta\left(\frac{n}{\sqrt{\log n}}\right)$. It is likely that Erdős made this stronger conjecture in a talk or in a paper listing open problems.

(5) Laszlo Székely [Székely (1993)] showed $g(n) = \Omega(n^{4/5})$. This proof is easier than the proofs in the papers by Chung and Chung *et al.* Moreover, it was a big breakthrough for techniques.

(6) Jozsef Solymosi and Casaba D. Toth [Solymosi and Toth (2001)] showed $g(n) = \Omega(n^{6/7})$ (Note that $6/7 = 0.857142857142\ldots$.) This was also a big breakthrough for techniques. Later papers were refinements of this method. In their book on combinatorial geometry, on page 116, Janos Pach and Micha Sharir [Pach and Sharir (2009)] state:

> *A close inspection of the Solymosi-Toth proof show that without any additional geometric ideas, it can never lead to a lower bound greater than $\Omega(n^{8/9})$.*

Note that $8/9 = 0.888\cdots$.

(7) Gabor Tardos [Tardos (2003)] showed that, for all $\epsilon > 0$, $g(n) = \Omega(n^{(4e/(5e-1))-\epsilon})$. Note that $\frac{4e}{5e-1} \approx 0.86353538281$. We paraphrase a comment they make on page 9:

> *The Ruzsa construction shows that one cannot use this technique to prove an $\Omega(n^{8/9})$ lower bound on*

*the number of distinct distances. A modification of
the proof of our theorem may help but the Ruzsa
construction still shows that, for all ϵ, a lower bound
of $\Omega(n^{(8/9)+\epsilon})$ is out or reach.*

They reference Ruzsa as *private communication*. I have not
been able to find a paper by Ruzsa with this result.

(8) Nets Katz and Gabor Tardos [Katz and Tardos (2004)]
showed that, for all $\epsilon > 0$, $g(n) = \Omega(n^{((48e-14e)/(55-16e))-\epsilon})$.
Note that $\frac{48-14e}{55-16e} \sim 0.86413751027$ (which is transcendental
so it's not going to have a repeating pattern).

As noted above Janos Pach, Micha Sharir, and Gabor Tardos
all said that $\Omega(n^{8/9})$ was as far as *the Solymosi-Toth techniques*
could go. I heard Solymosi say the same at a conference. The
work of Ruzsa formalizes this intuition. This *barrier result* did
not become an end in itself. There are no papers asking what
other open problems (interesting or not) could this barrier on
proof techniques be applied to.

So what did the community do? They found new techniques!

(1) Larry Guth and Nets Katz [Guth and Katz (2010)] solved
The Joints Conjecture: given any set of n lines in $\mathbb{R}^3$ there
are at most $O(n^{3/2})$ joints — points which are the intersection of three lines that are linearly independent (there
is an example of n lines with $\Omega(n^{3/2})$ joints). They used
techniques from algebraic geometry.

(2) Gyorgy Elekes and Micha Shari [Elekes and Sharir (2010)]
used the techniques of Larry Guth and Nets Katz to set up
a framework to study $g(n)$ and other problems. They solved
some problems with their framework, though they did not
get better lower bounds on $g(n)$.

(3) Larry Guth and Nets Katz [Guth and Katz (2015)] used the

framework of Gyorgy Elekes and Micha Sharir to obtain

$$g(n) = \Omega\left(\frac{n}{\log n}\right).$$

The first conjecture, $g(n) = \Omega(n^{1-\epsilon})$ is now solved. The second conjecture, $g(n) = \Omega\left(\frac{n}{\sqrt{\log n}}\right)$ is still open.

10.4 Comments From The Readers

My readers brought up a few examples outside of computer science where mathematicians proved barrier results:

(1) There is no quintic equation.
(2) CH is independent of Set Theory.
(3) The Parallel Postulate is independent of Geometry.

We explore each of these.

10.4.1 *There is No Formula for a Fifth Degree Equation*

Many years ago mathematicians had formulas for quadratic, cubic, and quartic equations. To be formal about it, the formulas were formed as follows. First we define $F_0, F_1, F_2, \ldots,$.

(1) F_0 is the set of coefficients.
(2) For all $i \geq 1$, F_i is obtained as follows:

  (a) Every element of F_i is in F_{i+1}.
  (b) If $a, b \in F_i$ then the following are in F_{i+1}: $a + b$, $a - b$, ab, a/b (if $b \neq 0$).
  (c) If $a \in F_i$ and $r \in \mathbb{Q}$ then all possible a^r are in F_{i+1}.

A *formula* is a combination of the F_i's which may depend on how the coefficient relate to each other. We give a familiar example.

The solutions to $ax^2 + bx + c$ are r_1, r_2 where

(1) If $a = 0$ and $b = 0$ and $c = 0$ then all complex numbers are roots.
(2) If $a = 0$ and $b = 0$ and $c \neq 0$ then there are no roots.
(3) If $a = 0$ and $b \neq 0$ then $r_1 = r_2 = \frac{-c}{b}$.
(4) If $a \neq 0$ and $b^2 - 4ac = 0$ then $r_1 = r_2 = \frac{-b}{2a}$,
(5) If $a \neq 0$ and $b^2 - 4ac \neq 0$ then $r_1 = \frac{-b+\sqrt{b^2-4ac}}{2a}$ and $r_2 = \frac{-b-\sqrt{b^2-4ac}}{2a}$.

Mathematicians were unable to obtain a formula for a fifth degree polynomial. Abel and Galois (independently) showed that *there was no such formula.* Okay, so what to do?

Change your Goals

Let us allow one more operation in our definition of the F_i.

(1) If $a \in F_i$ then let the solutions to $x^5 + x + a$ be in F_{i+1}.

George Jerrard, in 1835, showed that by using these new formulas any quintic could be solved.
This can be interpreted as

(1) Galois proved a barrier result.
(2) Jerrard found a way around it. But he changed the problem. Is that cheating?

What about the sixth degree equation? There are results but they are complicated. What about the seventh degree equation? This question is related to Hilbert's thirteenth problem, and is partially solved. What about the eighth degree equation? I have not been able to find if there is a solution using other functions. I leave the search for such as an exercise for the reader.

10.4.2 *A Personal Digression*

Looking up and revisiting the unsolvability of the quintic, and learning Jerrard's result (which I did not know before I wrote the blog, and looked into readers' comments) reminded me of why I studied mathematics in the first place. In ninth grade, when my teacher said *there is a quadratic equation, a cubic equation, a quartic equation, but not a quintic equation.* I decided right then and there to study math in college and learn why. When I applied to colleges in 1975, the following was part of my personal statement:

> *I want to be a math major so I can find out why there is no algebraic solution for the roots of a quintic equation, or any general polynomial equation of degree greater than four, in terms of explicit algebraic operations.*

I do not know if that essay helped or hurt my college admissions choices; fortunately they did not realize that it was a direct quote from the Wikipedia entry on Abel.

At SUNY Stonybrook I learned the proof of the unsolvability of the quintic in my junior year. Fortunately by then I had picked up other interests within math. In particular I went into computability because of the unsolvability of the Halting problem.

10.4.3 *The Independence of The Continuum Hypothesis (CH) From Set Theory*

Definition 10.1. Let A be a set of axioms and f is a statement in mathematics. Assume we have a reasonable set of rules of inference.

(1) $A \vdash f$ means that from A one can prove f.

(2) $A \nvdash f$ means that from A one cannot prove f. Note that f might still be true. It may be that A was too weak a system to prove f. (We will use $\vdash$ and $\nvdash$ in the next subsection also.)

(3) ZFC is a set of axioms and rules of inference from which one can derive virtually all theorems of mathematics. Most of the exceptions are in Logic. ZFC stands for Zermelo-Frankel with The Axiom of Choice.

(4) A *model of set theory* is a model where ZFC holds.

(5) The first of Hilbert's 23 problems was to show that there is no infinity between $\mathbb{N}$ and $\mathbb{R}$. This is known as *The Continuum Hypothesis* and denoted *CH*.

Gödel showed that ZFC $\nvdash \neg$CH by giving a model of set theory where CH was true. Cohen showed that ZFC $\nvdash$ CH by giving a model of set theory where $\neg$CH was true. Hence CH is *independent of set theory*.

Is the independence of CH a barrier result? We present both viewpoints.

YES. CH is either true or false but ZFC is not up to the task of finding it. Perhaps if we add some other axioms we can resolve CH. Large Cardinal Axioms and the Axiom of Determinacy are popular choices. But (1) none of these so far have resolved CH, and (2) none of these can be considered so obvious as to be an axiom.

NO. CH has no answer. There are different models of set theory. In some CH is true and in some CH is false. No model has a reason to be called *the right one*. So what to do? We return to this question after the next section which has a similar conundrum.

10.4.4 *The Independence of The Parallel Postulate from Geometry*

Warning: This section is somewhat informal. In particular, Euclidean Geometry was refined by Hilbert to be *more rigorous*. This is not important for our exposition.

Euclid stated five axioms from which he derived all of his theorems. The first four were quite reasonable. We denote these E1-4. A *model of geometry* is a model where E1-4 hold. The fifth axiom, which we denote E5, is (in modern terminology):

> *If p is a point and L is a line that does not contain p then there is exactly one line through p that is parallel to L.*

The original version from Euclid is awkward. We quote it from Wolfram's MathWorld.

> *If two lines are drawn which intersect a third in such a way that the sum of the inner angles on one side is less than the sum of two right angles, then the two lines inevitably must intersect each other on that side if extended far enough.*

Mathematicians tried to show that E5 could be derived from the other four axioms. Why? Perhaps because the original version of E5 is awkward and un-intuitive. One approach was to see what could be derived from negating E5.

If $E5$ is false then one of the following is true:

(1) (RA, Riemann's Axiom. Nobody calls it that.) There is a point p and a line L, with p not on L, such that there are *no* lines through p that are parallel to L.

(2) (LA, Lobachevsky-Bolyai's Axiom. Nobody calls it that.) There is a point p and a line L, with p not on L, such that there are *at least two* lines through p that are parallel to L.

Riemann showed E1-4 $\not\vdash$ E5 by giving a model of geometry where RA is true. We call this *Riemannian Geometry*. Lobachevsky and Bolyai (independently, though see Tom Lehrer's song *Lobaschevsky* for an alternative viewpoint) showed E1-4 $\not\vdash$ E5 by giving a model of geometry where *LA* is true. We call this *Hyperbolic Geometry*. It is well known that there is a model of geometry where E5 is true. We call this *Euclidean Geometry*. Hence E5 is *independent E1-4*.

Is the independence of E5 a barrier result? We present both viewpoints.

YES. Nobody believes this. See the NO answer for why.

NO. There are three different geometries (actually there are far more but we ignore that for now). Which one is right? That's the wrong question! The proper attitude is

Use the Right Tool for the Right Job!

A bridge builder uses in Euclidean geometry. A crystallographer uses hyperbolic geometry. An astrophysicist uses Riemannian geometry.

There may be a lesson here for CH. In the future it may be that some real world phenomena is best modeled with a set theory where CH is true, and some other real world phenomenon is best modeled with a set theory where CH is false. If that happens then, much like in geometry, there will be no "right model."

References

Baker, T., Gill, J., and Solovay, R. (1975). Relativizations of the $P =?$ NP question, *SIAM Journal on Computing* **4**, pp. 431–442.

Buss, J. F. (1987). A theory of oracle machines, in *Proceedings of the 2nd IEEE Conference on Structure in Complexity Theory*, Cornell NY (IEEE Computer Society Press), pp. 175–181.

Chung, F. (1984). The number of different distances determined by n points in the plane, *Journal of Combinatorial Theory, Series A* **36**, pp. 342–354, http://math.ucsd.edu/~fan/ or http://www.cs.umd.edu/~gasarch/erdos_dist/erdos_dist.html.

Chung, F., Szemerédi, E., and Trotter, W. (1992). The number of different distances determined by a set of points in the Euclidean plane, *Discrete & Computational Geometry* **7**, pp. 1–11, http://math.ucsd.edu/~fan/ or http://www.cs.umd.edu/~gasarch/erdos_dist/erdos_dist.html.

Elekes, G. and Sharir, M. (2010). Incidences in three dimensions and distinct distances in the plane, in *Proceedings of the 26th annual Symposiusm on computational geometry*, pp. 413–422.

Erdős, P. (1946). On sets of distances of n points, *The American Mathematical Monthly* **53**, pp. 248–250, http://www.cs.umd.edu/~gasarch/erdos_dist/erdos_dist.html or http://www.jstor.org/.

Gariabldi, J., Iosevich, A., and Senger, S. (2010). *The Erdős distance problem* (American Mathematics Society).

Gasarch, W. (1987). Oracles for deterministic versus alternating classes, *SIAM Journal on Computing* **16**, pp. 613–627.

Guth, L. and Katz, N. H. (2010). Algebraic methods in discrete analogs of the Kakeya problem, *Advances in Mathematics* **225**, pp. 2828–2839.

Guth, L. and Katz, N. H. (2015). On the erdos distinct distances problem in the plane, *Annals of Mathematics* **181**, pp. 155–190.

Solymosi, J. and Toth, C. D. (2001). Distinct distances in the plane, *Discrete & Computational Geometry* **25**, pp. 629–634, http://www.cs.umd.edu/~gasarch/erdos_dist/erdos_dist.html or http://citeseer.ist.psu.edu/.

Katz, N. and Tardos, G. (2004). A new entropy inequality for the Erdös distance problem, in *Towards a theory of Geometric Graphs, Contemporary Mathematics*, Vol. 342 (American Mathematical Society, Providence, Rhode Island), pp. 119–126, http://www.renyi.hu/~tardos/ or http://www.cs.umd.edu/~gasarch/erdos_dist/erdos_dist.html.

Ladner, R. and Lynch, N. (1975). Relativiation of questions about log space computability, *Theoretical Computer Science* **1**, pp. 103–123.

Moser, L. (1952). On the different distances determined by *n* points, *The American Mathematical Monthly* **59**, pp. 85–91, http://www.cs.umd.edu/~gasarch/erdos_dist/erdos_dist.html or http://www.jstor.org/.

Pach, J. and Sharir, M. (2009). *Combinatorial geometry and its algorithmic applications* (American Mathematical Society), book 152 in a series called *Mathematical surveys and Monographs*.

Razborov, A. A. and Rudich, S. (1997). Natural proofs, *Journal of Computer and System Sciences* **55**, 1, pp. 24–35, prior version in *ACM* Sym on Theory of Computing, 1994 (STOC).

Ruzzo, W., Simon, J., and Tompa, M. (1982). Space-bounded hierarchies and probabilistic computations, *Journal of Computer and System Sciences* **28**, pp. 216–230.

Székely, L. (1993). Crossing numbers and hard Erdős problems in discrete geometry, *Combinatorics, Probability and Computing* **11**, pp. 1–10, http://www.cs.umd.edu/~gasarch/erdos_dist/erdos_dist.html.

Tardos, G. (2003). On distinct distances and distinct distances, *Advances in Mathematics* **180**, pp. 275–289, http://www.renyi.hu/~tardos/ or http://www.cs.umd.edu/~gasarch/erdos_dist/erdos_dist.html.

PART 2

Problems with a Point

Chapter 11

Funny Answers to Math Questions

Prior Knowledge Needed: None.

11.1 Point

If you were taking a math test and didn't know how to solve a question: what would you do? Think harder? Think deeper? Maybe it will come to you. You will try since you care about your grade.

What if you are taking an exam in a Math Competition? If you get a 0 nothing bad happens. It might be easier to give up. And what do you do if you give up? Perhaps write some funny answers.

Two problems from the Maryland Math Competition stand out as having gotten funny answers. We state the problems, the funny answers, and the real solutions.

11.2 The University of Maryland High School Mathematics Competition

Every year, the University of Maryland runs a math competition for high school students [Maryland Mathematics Competition] in two parts. In the first part, the students take a 75 minute multiple choice test of 25 questions, five answers for each question: +4 for a right answer, -2 for a wrong answer. It is a hard exam, with scores typically ranging from -40 up to 100. Around 400 students take the first part.

The students who get over a threshold on Part I are invited to take Part II, which is a 2-hour exam with five math problems, 30 points each. These are serious problems of the type you would see in a Putnam exam, though easier. We try to make the problems start out easy and get harder. The winners are determined by adding their Part I and Part II scores. Around 220 students take Part II.

11.3   Cans of Paint

In the year 2000 I made up and graded the following Part II problem for the Maryland Math Competition.

There are 2000 cans of paint. Show that at least one of the following two statements is true:

- *There are at least 45 cans of the same color.*
- *There are at least 45 cans that are different colors.*

It was Problem 1 so it was supposed to be easy. In fact, 95% of the students got it right. Note that the students taking Part II did well on Part I, so they are good students.

There were two funny answers, coming to opposite conclusions.

Funny Answer One:
Paint cans are grey. Hence they are all the same color. Therefore there are 2000 cans that are the same color.

Funny Answer Two:
If you look at a paint color really really carefully there will be differences. Hence, even if two cans seem to both be (say) RED, they are really different. Therefore there are 2000 cans of different colors.

Were they serious? The first one points to a problem with the phrasing of the question — I clearly did not mean the cans themselves, and all of the other students knew that, but looking at the problem it could be interpreted that way. This person might have been serious. Perhaps I should have graded that person right and everyone else wrong.

The second one I can't imagine was serious.

11.4 Points in the Plane

I was assigned to grade the following problem from the Maryland Math Competition from 2007 (for High School Students):

Let ABC be a fixed triangle. Let COL be any 2-coloring of the plane where each point is colored with red or green. Prove that there is a triangle DEF in the plane such that DEF is similar to ABC and the vertices of DEF all have the same color.

I think I was assigned to grade it since it looks like the kind of problem I would make up, even though I didn't. It was problem 5 (out of 5), so it was the hardest. About 100 students of the students tried it; 5 got full credit, and 8 got partial credit (and they didn't get much).

There were two funny answers:

Funny Answer One:
All the vertices are red because I can make them whatever color I want. I can also write at a 30 degree angle to the bottom of this paper if thats what I feel like doing at the moment. [The student's answer was written at a 30 degree angle to the bottom of the paper.] Just like $2 + 2 = 5$ if thats what my math teacher says. Math is pretty subjective anyway.

Funny Answer Two:

I like to think that we live in a world where points are not judged by their color, but by the content of their character. Color should be irrelevant in the plane. To prove that there exists a group of points where only one color is acceptable is a reprehensible act of bigotry and discrimination.

Were they serious? Hard to say, but I'm impressed that the second student knows Martin Luther King's *I have a dream* speech.

11.5   The Real Answers

Cans of Paint

There are 2000 cans of paint. Show that at least one of the following two statements is true:

- *There are at least 45 cans of the same color.*
- *There are at least 45 cans that are different colors.*

If there are 45 different colors of paint then we are done. Assume there are ≤ 44 different colors. If all colors appear ≤ 44 times then there are $44 \times 44 = 1936 < 2000$ cans of paint, a contradiction.

Points in the Plane

Each point in the plane is colored either red or green. Let ABC be a fixed triangle. Prove that there is a triangle DEF in the plane such that DEF is similar to ABC and the vertices of DEF all have the same color.

Fix a 2-coloring of the plane.

Claim 1: There are three monochromatic points on the x-axis that are equally spaced.

Proof: Clearly there are two (actually infinitely many) points on the x-axis that are the same color. We call them p_1 and p_2 and assume they are RED. Refer to Figure 11.1.

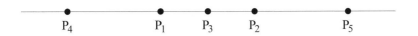

Fig. 11.1   There are Three Equally Spaced Points of the Same Color.

If p_3, the midpoint of p_1, p_2, is RED then p_1, p_3, p_2 are all RED and are the desired points. Hence we assume p_3 is GREEN.

Let p_4 be a point on the x-axis such that $|p_1 - p_4| = |p_2 - p_1|$. If p_4 is RED then p_4, p_1, p_2 are all RED and are the desired points. Hence we assume p_4 is GREEN.

Let p_5 be the other point (not p_4) on the x-axis such that $|p_5 - p_2| = |p_2 - p_1|$. If p_5 is RED then p_1, p_2, p_5 are all RED and are the desired points. Hence we assume p_5 is GREEN.

In the only case left p_3, p_4, p_5 are all GREEN and are the desired points.

End of Proof of Claim 1

Claim 2: There are three monochromatic points that form a triangle similar to ABC.

Proof: Let P, Q, R be three monochromatic equally spaced points on the x-axis (that we know exist from Claim 1). Assume they are all RED. Refer to Figure 11.2.

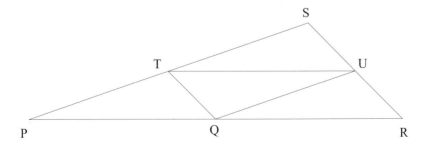

Fig. 11.2   Triangle Similar to ABC with Monochromatic Vertices.

Pick S such that PRS is similar to ABC. Let T be the midpoint of PS and U be the midpoint of RS. The two new triangles PQT and QRU are also similar to ABC.

If S is RED then PRS is monochromatic. If T is RED then PQT is monochromatic. If U is RED then QRU is monochromatic. In all these cases we have a monochromatic triangle similar to ABC. Otherwise TUS is monochromatic (all GREEN vertices) and similar to ABC.

End of Proof of Claim 2

References

Maryland Mathematics Competition (1998–). http://www.math.umd.edu/highschool/mathcomp/.

Chapter 12

What is an Elegant Proof?
Some Coloring Theorems

Prior Knowledge Needed: None.

12.1 The Point

We will present a sequence of proofs and discuss if they are elegant. Elegance is not a rigorous concept; however, discussing it is enlightening.

12.2 The Problems

Notation 12.1. If $n \in \mathbb{N}$ then $[n]$ is the set $\{1, \ldots, n\}$.

Definition 12.1. A *proper c-coloring of* $[n]$ is a c-coloring of $[n]$ so that there is no monochromatic pair a square apart. Let $q(c)$ be the least number so that $[q(c)]$ is not properly c-colorable.

Example 12.1.

(1) The following is a proper 3-coloring of $[9]$ using colors RED, BLUE, and GREEN. The reader can check that for all $1 \leq x \leq 8$, $COL(x) \neq COL(x + 1)$ and for all $1 \leq x \leq 5$, $COL(x) \neq COL(x + 4)$.

1	2	3	4	5	6	7	8	9
R	B	R	B	G	R	B	R	B

(2) The following is a non-proper 3-coloring of [9] using colors RED, BLUE, and GREEN. The reader can find the x such that $COL(x) = COL(x + 4)$.

1	2	3	4	5	6	7	8	9
R	B	R	B	G	B	G	R	B

Consider the following challenges.

(1) Show that $q(2)$ exists.
(2) Find $q(2)$.
(3) Show that $q(3)$ exists.
(4) Find $q(3)$.
(5) Show that $q(4)$ exists.
(6) Find $q(4)$.
(7) Show that, for all $c \geq 5$, $q(c)$ exists.
(8) Find, for all $c \geq 5$, $q(c)$.

We will solve 1 through 5.

12.3   The $c = 2$ Case

Theorem 12.1. $q(2) = 5$

Proof. Note that 1 is a square. No matter how you color $\{1, 2, 3, 4, 5\}$ either there will be two adjacent numbers the same color, or 1 and 4 are the same color. In either case there are two numbers a square apart that are the same color. Hence $q(2) \leq 5$.

The following is a proper 2-coloring of $\{1, 2, 3, 4\}$, proving that $q(2) \geq 5$.

1	2	3	4
R	B	R	B

□

I would call this proof elegant; however, I've heard people say the proof is too easy to be elegant. To be fair, they are wrong, and I am right.

Some of my readers commented *OH, 1 is a square? I thought that was cheating.* I leave it to the reader to give an elegant proof of the following: For any n_0 there exists N such that any 2-coloring of $\{1, \ldots, N\}$ has a monochromatic pair a square apart where that square is $\geq n_0^2$.

12.4 The $c = 3$ Case

Theorem 12.2.

(1) $q(3) \leq 1157$
(2) $q(3) \leq 68$

Proof. 1) Let COL be a 3-coloring of $\{1, \ldots, 1157\}$. Figure 12.1 shows that, either there are two numbers a square apart that are the same color or, for all x, $COL(x) = COL(x + 34)$. In the later case we have

$$COL(1) = COL(1+34) = COL(1+2\times34) = \cdots = COL(1+34^2)$$

hence there are two numbers a square apart that are the same color.

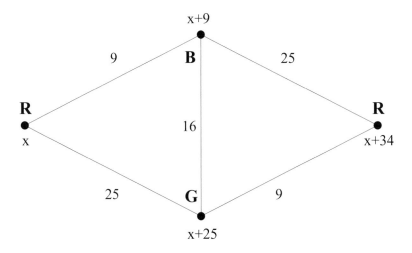

Fig. 12.1 $COL(x) = COL(x + 34)$.

2) Let COL be a 3-coloring of $\{1, \ldots, 68\}$. Figure 12.2 shows that, either there are two numbers a square apart that are the same color or, for all $x \geq 10$, $COL(x) = COL(x+7)$. In the later case we have

$$COL(10) = COL(10+7) = COL(10+2\times 7) = \cdots = COL(1+7^2)$$

hence there are two numbers a square apart that are the same color.

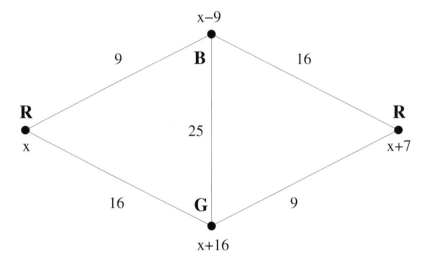

Fig. 12.2 $COL(x) = COL(x+7)$.

□

Both of the proofs in Theorem 12.2 are elegant in that they were short and elementary. Can we obtain a smaller value of N. Yes. We will show that $N = 29$ is the smallest possible value of N in the next section.

The following was problem five on the *The University of Maryland High School Mathematics Competition* in 2006. (For information on this competition see Chapter 11.)

Show that for any 3-coloring of $\{1, \ldots, 2006\}$ there exist two distinct numbers that are a square apart and the same color.

This was the last problem and meant to be difficult. Around 240 students took the Part II exam. Around 40 students tried this problem. Of those, only 10 got it right.

12.5 Exact Bounds for the $c = 3$ Case

The results in this section were proven by both Matt Jordan and myself. Matt was an undergraduate working with me.

12.6 The $c = 4$ Case

Theorem 12.3. $q(3) = 29$.

Proof. Let COL be any 3-coloring of $[29]$. We know from Section 12.4 that

$$(\forall x)[10 \leq x \leq 13 \implies \text{COL}(x) = \text{COL}(x + 7)].$$

We refer to this fact as FORCE.

We can assume, without loss of generality, that $\text{COL}(10) = $ R. Since $11 - 10 = 1^2$ we know that $\text{COL}(11) \neq$ R. We can, without loss of generality, assume that $\text{COL}(11) = $ B.

17: By FORCE $\text{COL}(17) = \text{COL}(10) = $ R

18: By FORCE $\text{COL}(18) = \text{COL}(11) = $ B.

10	11	12	13	14	15	16	17	18	19	20
R	B						R	B		

19: Since $\text{COL}(10) = $ R and $\text{COL}(18) = $ B, $\text{COL}(19) = $ G.

12: By FORCE $\text{COL}(12) = \text{COL}(19) = $ G.

10	11	12	13	14	15	16	17	18	19	20
R	B	G					R	B	G	

20: Since $\text{COL}(11) = $ B and $\text{COL}(19) = $ G, $\text{COL}(20) = $ R.

13: By FORCE $\text{COL}(13) = \text{COL}(20) = $ R.

10	11	12	13	14	15	16	17	18	19	20
R	B	G	R				R	B	G	R

Now we have that $COL(17) = COL(13) = R$. But $17 - 13 = 2^2$. This is a contradiction.

We present a proper 3-coloring of $[28]$:

1	2	3	4	5	6	7	8	9	10	11	12	13	14
G	R	B	G	R	B	R	B	G	R	B	G	B	G

15	16	17	18	19	20	21	22	23	24	25	26	27	28
R	B	R	B	G	R	B	R	B	G	R	B	G	R

$\square$

Is the above proof elegant? That's a matter of taste. It's elementary but involves a long chain of reasoning.

12.7 The $c = 4$ Case

Does $q(4)$ exist? If so then what is it? How about a bound on it? Yes, $q(4)$ exists! One can show this by the polynomial van der Warden theorem (which we state in the next section). Such a proof is rather advanced and gives good bounds. It's elegant — to a Ramsey Theorist.

Is there an elegant proof (as there is for $q(2)$ and $q(3)$)? I was tempted to find out by putting the following in the 2007 competition:

Show that for any 4-coloring of $\{1, \ldots, 2007\}$ *there exists two distinct numbers that are a square apart and the same color.*

The committee in charge of the exam (wisely) talked me out of it.

How about a computer proof? In the summer of 2015 I had students Burcu Chanackci, Hannah Christenson, Robert Fleishman, Nicole McNabb, and Daniel Smolyak work on finding Ramsey Numbers using Sat Solvers. Their program found that $q(4) = 58$. The good news is that we know the answer! The bad news is that *a computer told me the answer*, so the proof is not elegant.

In the fall of 2018 I had a student Zach Price work on getting a more elegant proof.

Good news: He found an elementary proof that $q(4)$ exists. Yeah!

Bad news: The bounds on $q(4)$ are enormous.

Good news: Once you have the proof you can easily verify it. So Zach can *claim* he found it by hand. I doubt anyone would believe that.

Bad news: It is unlikely that a program will find bounds for $q(5)$ since almost surely the numbers involved will be ginormous.

Good news: The bounds on $q(4)$ are still much smaller than one would get with the poly van der Warden theorem.

Here is his proof:

Theorem 12.4. *Let* $M = 290,085,289$. $q(4) \leq N^2 + 1 = 84,149,474,894,213,522$.

Proof. Let COL be a 4-coloring of $\{1, \ldots, N^2+1\}$. Figure 12.3 shows that, either there are two numbers a square apart that are the same color or, for all x, $COL(x) = COL(x + N)$. In the later case we have

$$COL(1) = COL(1 + N) = COL(1 + 2N) = \cdots = COL(1 + N^2)$$

hence there are two numbers a square apart that are the same color.

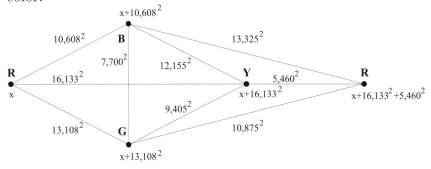

Fig. 12.3 $COL(x) = COL(x + N)$. □

We compare the three proofs which we call pvdw (poly van der Warden), sat (used sat solvers) and Zach (Theorem 12.4). The pvdw proof uses interesting math but is rather advanced and gives terrible bounds. The sat proof gives exact bounds but we do not gain insights and cannot be verified. The Zach proof gives insight, and can be verified, but the bounds are large (though not even close to those from pvdw).

Which do you prefer? In the interest of full disclosure I admit that I like Zach's proof the best. Even so, I won't be putting the problem on a High School Math Competition.

12.8   Formal Statement of the Polynomial van Der Warden Theorem

Vitaly Bergelson and Alexander Leibman [Bergelson and Leibman (1996)] gave a proof of the Polynomial van der Warden Theorem that used advanced mathematics. Mark Walters [Walters (2000)] gave an elementary proof; however, it is still difficult.

We first state the ordinary van der Warden Theorem for contrast.

Notation 12.2. If $W \in \mathbb{N}$ then $[W] = \{1, \ldots, W\}$.

Van der Warden proved the theorem that bears his name in 1927.

Theorem 12.5. *Let $k, c \in \mathbb{N}$. There exists $W = W(k, c)$ such that for all c-colorings of $[W]$ there exists a, d such that*

$$a, a + d, a + 2d, \cdots, a + (k - 1)d$$

are all the same color.

To generalize this theorem one ponders what to replace

$$d, 2d, 3d, \ldots, (k - 1)d$$

with. The answer: polynomials in $\mathbb{Z}[x]$ with 0 constant term.

This is the Polynomial van der Warden Theorem:

Theorem 12.6. *Let* $p_1, \ldots, p_k \in \mathbb{Z}[x]$ *such that* $(\forall i)[p_i(0) = 0]$. *Let* $c \in \mathbb{N}$. *There exists* $W = W(p_1, \ldots, p_k; c)$ *such that for all c-colorings of* $[W]$ *there exists* a, d *such that*

$$a, a + p_1(d), a + p_2(d), \cdots, a + p_k(d)$$

are all the same color.

We have just been using the case where $k = 1$ and $p_1(x) = x^2$.

References

Bergelson, V. and Leibman, A. (1996). Polynomial extensions of van der Warden's and Szemerédi's theorems, *Journal of the American Mathematical Society* **9**, pp. 725–753, http://www.math.ohio-state.edu/~vitaly/ or http://www.cs.umd.edu/~gasarch/vdw/vdw.html.

Walters, M. (2000). Combinatorial proofs of the polynomial van der Warden theorem and the polynomial Hales-Jewett theorem, *Journal of the London Mathematical Society* **61**, pp. 1–12, http://jlms.oxfordjournals.org/cgi/reprint/61/1/1 or http://jlms.oxfordjournals.org/ or http://www.cs.umd.edu/~gasarch/vdw/vdw.html.

Chapter 13

A High School Math Competition Problem Inspires an Infinite Number of Proofs

With Help From
Vince Cozzo, Carolyn Gasarch, Dilhan Salgado

Prior Knowledge Needed: None, but you have to care about reciprocals.

13.1 The Point

The following was problem number 2 (out of 5) on Part II of the *The University of Maryland High School Mathematics Competition* in 2010. (For information on this competition see the Chapter 11.)

(a) The equations $\frac{1}{2} + \frac{1}{3} + \frac{1}{6} = 1$ and $\frac{1}{2} + \frac{1}{3} + \frac{1}{7} + \frac{1}{42} = 1$ express 1 as the sum of three (respectively four) distinct positive integers. Find five distinct positive integers $a < b < c < d < e$ such that $\frac{1}{a} + \frac{1}{b} + \frac{1}{c} + \frac{1}{d} + \frac{1}{e} = 1$.

(b) Prove that for any integer $m \geq 3$ there exists m positive integers $d_1 < d_2 < \cdots < d_m$ such that $\frac{1}{d_1} + \cdots + \frac{1}{d_m} = 1$.

Of the 200 students who took Part II, 188 attempted this problem. They all got Part (a) correct. We list all of those answers in the Section 13.10. For Part (b), 160 got it correct. There were four different correct solutions. Since there were four proofs, I wondered *if there are more?* The answer is yes and YES! It turns out, there are an infinite number of proofs. This chapter will describe our quest to find them. Many points are made along the way.

In Section 13.3 we present the four proofs the students gave, which we call SOL1, SOL2, SOL3, SOL4. In Sections 13.4, 13.5, and 13.6 we give an infinite number of proofs based on SOL1. In Section 13.7 we give another infinite number of proofs that are not a variant of what the students did. In Section 13.8 we give some empirical evidence about how we can improve our theorems. In Section 13.9 we pout about the fact that the story told here, with history and context, would have to be removed from the journal version. In Section 13.10, we give a table of information about what the students did.

13.2   Notation and Helpful Lemma

Definition 13.1. An *Egyptian Fraction* is a fraction of the form $\frac{1}{m}$ where $m \in \mathbb{N}$.

Definition 13.2.

(1) A *nice sequence of natural numbers* is a sequence whose reciprocals sum to 1.
(2) A *nice n-sequence of natural numbers* is a nice sequence of length n.

The following is well known.

Lemma 13.1.

(1) Every positive rational number is the sum of distinct Egyptian fractions.

(2) If $m \in \mathbb{N}$ then there exists a nice sequence where every element is divisible by m. (This follows from Item 1 since you can multiply $\sum_{i=1}^{m} \frac{1}{a_i} = m$ by $\frac{1}{m}$.)

13.3 Four Correct Solutions and An Interesting Incorrect Solution

Notation 13.1. Let $P(n)$ be the statement: *There exists a nice n-sequence.*

Theorem 13.1. *Then* $(\forall n \geq 3)[P(n)]$.

Proof. We sketch the four correct solutions submitted and an interesting incorrect one. All are by induction on n.

Let Base3 and Base4 be the following equations:

$$\tfrac{1}{2} + \tfrac{1}{3} + \tfrac{1}{6} = 1$$

$$\tfrac{1}{2} + \tfrac{1}{3} + \tfrac{1}{7} + \tfrac{1}{42} = 1$$

They will be used as base cases.

SOLUTION ONE: Base3 is the base case. For $n \geq 4$ use

$$\frac{1}{d} = \frac{1}{d+1} + \frac{1}{d(d+1)}$$

to go from $P(n-1)$ to $P(n)$.

133 students submitted this solution which was by far the most common. This may be because the exam gave them base3 and base4, and this is the only proof of the four that produces base4 from base3.

If you apply the induction step to $d = 3$ twice you get (2,3,7,43,1806) for Part (a). Only 91 students submitted (2,3,7,43,1806). And of those, 74 submitted SOLUTION ONE for Part (b).

One of the students who submitted this solution to Part (b), but did not submit (2,3,7,43,1806) for Part (a) told me that he found another solution for Part (a) to avoid having to multiply 42 by 43.

SOLUTION TWO: Base3 and Base4 are the base cases. Use

$$\frac{1}{d} = \frac{1}{2d} + \frac{1}{3d} + \frac{1}{6d}$$

to go from $P(n-1)$ to $P(n+1)$.

This was the second most common solution with 21 students submitting it. This leads to (2,3,12,18,36) for Part (a). Only twelve students submitted (2,3,12,18,36) for Part (a). And not all of those twelve students submitted SOLUTION TWO.

SOLUTION THREE: Base3 is the base case. Load the induction hypothesis with the additional assumption that d_n is even.

Use

$$\frac{1}{d} = \frac{1}{(3d/2)} + \frac{1}{3d}$$

to go from $P(n-1)$ to $P(n)$.

This was the third most common solution with four students submitting it. This leads to (2,3,7,63,126) for Part (a). Six students submitted it for Part (a). Including all four who submitted SOLUTION THREE for Part (b).

SOLUTION FOUR: Base3 is the base case.

Use

$$\frac{1}{d_1} + \cdots + \frac{1}{d_{n-1}} = 1 \implies \frac{1}{2} + \frac{1}{2d_1} + \cdots + \frac{1}{2d_{n-1}} = 1$$

to go from $P(n-1)$ to $P(n)$.

This was the fourth most common solution (or the least most common solution) with two students submitting it. It leads to (2,4,6,14,84) for Part (a). Three students submitted it for Part (a) including the two who submitted SOLUTION FOUR for Part (b).

ALMOST SOLUTION: Here is a proof that seems to not require induction.

Assume that, for all n, there is such an A, such that A is the sum of n of its distinct divisors. For example, if $n = 3$ then $A = 36 = 6 + 12 + 18$ works. (Given n, is there always such an

A? Thats the problem with the proof.) Let $d_1, \ldots, d_n$ be the divisors of A that sum to A. Consider

$$\frac{d_1}{A} + \cdots + \frac{d_n}{A}.$$

Since, for all i, d_i divides A, this is a sum of distinct reciprocals. Since $\sum_{i=1}^{n} d_i = A$, the sum of these reciprocals is 1.

Could this solution work? The premise that it needs, for all n there is such an A that is the sum of n of its distinct divisors, is true! How do we prove it? Alas, the only proof we know uses the theorem we are trying to prove. Oh well. We leave it to the reader to prove the implication. □

13.4   SOLUTION ONE-1 and SOLUTION ONE-2

I had an undergraduate, Vince Cozzo, write a program that, given n, B, generated the first B solutions in lexicographic order to

$$\frac{1}{d_1} + \cdots + \frac{1}{d_n} = 1.$$

The idea was to see if we could generate the four proofs of Theorem 13.1. The first four solutions are:

$$\tfrac{1}{2} + \tfrac{1}{3} + \tfrac{1}{6} = 1$$

$$\tfrac{1}{2} + \tfrac{1}{3} + \tfrac{1}{7} + \tfrac{1}{42} = 1$$

$$\tfrac{1}{2} + \tfrac{1}{3} + \tfrac{1}{7} + \tfrac{1}{43} + \tfrac{1}{1806} = 1$$

$$\tfrac{1}{2} + \tfrac{1}{3} + \tfrac{1}{7} + \tfrac{1}{43} + \tfrac{1}{1807} + \tfrac{1}{3263442} = 1$$

These lead to SOLUTION ONE. Even so, with the benefit of hindsight we can rewrite the proof in a way that will fit a pattern. We call it SOLUTION FIVE-1 for reasons that will become clear. We will write it in a odd way so that it fits into a pattern.

SOLUTION ONE-1: We prove $(\forall n \geq 3)[P(n)]$. Base3 is the base case. Let $f_1(x) = \frac{x(x-1)}{1}$. Load the induction hypothesis with the additional assumptions that

- $d_{n-1} \equiv 0 \pmod{1}$ (this is always true).
- $d_{n-2} < d_{n-1}$ (this is always true)
- $d_n = f(d_{n-1})$.
- $d_{n-1} \geq 2$.

Use

$$\frac{1}{d_{n-2}} + \frac{1}{d_{n-1}} + \frac{1}{f_1(d_{n-1})} =$$

$$\frac{1}{d_{n-2}} + \frac{1}{d_{n-1}} + \frac{1}{d_{n-1}(d_{n-1}-1)+1} + \frac{1}{f_1(d_{n-1}(d_{n-1}-1)+1)}$$

to go from $P(n-1)$ to $P(n)$. Use $d_{n-1} \geq 2$ to show

$$d_{n-1} < d_{n-1}(d_{n-1}-1)+1$$

We now produce SOLUTION ONE-2:

SOLUTION ONE-2: We prove $(\forall n \geq 4)[P(n)]$. Base4 is the base case. Let $f_2(x) = \frac{x(x-2)}{2}$. Load the induction hypothesis with the additional assumptions that

- $d_{n-1} \equiv 0 \pmod{2}$
- $d_{n-2} < d_{n-1} - 1$
- $d_n = f(d_{n-1})$.
- $d_{n-1} \geq 3$.

Use

$$\frac{1}{d_{n-2}} + \frac{1}{d_{n-1}} + \frac{1}{f_2(d_{n-1})} = \frac{1}{d_{n-2}} + \frac{1}{d_{n-1} - 1}$$

$$+ \frac{1}{(d_{n-1}-1)(d_{n-1}-2)+2} + \frac{1}{f_2((d_{n-1}-1)(d_{n-1}-2)+2)}$$

to go from $P(n-1)$ to $P(n)$. Use $d_{n-1} \geq 3$ to prove

$$d_{n-1} - 1 < (d_{n-1} - 1)(d_{n-1} - 2) + 2.$$

Note 13.1. We could not have used $\frac{1}{2} + \frac{1}{3} + \frac{1}{6} = 1$ for the base case since it does not satisfy the induction hypothesis. When we generalize to get SOLUTION-ONE-a we will need to be careful to get a base case that satisfies the loaded induction hypothesis.

13.5   For all a, SOLUTION ONE-a

Definition 13.3. Let $a \in \mathbb{N}$. Let $f_a(x) = \frac{x(x-a)}{a}$.

Lemma 13.2. *Let $a \geq 1$ and $b, d, x \in \mathbb{N}$.*

(1)

$$\frac{1}{d} + \frac{1}{f_a(d)} = \frac{1}{d-a}$$

(2)

$$\frac{1}{x} + \frac{1}{x(x-1)+a} + \frac{1}{f_a(x(x-1)+a)} = \frac{1}{x-1}$$

(3)

$$\frac{1}{d-a+1} + \frac{1}{(d-a+1)(d-a)+a}$$

$$+ \frac{1}{f_a((d-a+1)(d-a)+a)} = \frac{1}{d-a}$$

(This follows from item 2 with $x = d - a + 1$.)

(4)

$$\frac{1}{b} = \frac{1}{b+1} + \frac{1}{b(b+1)+a} + \frac{1}{f_a(b(b+1)+a)}$$

(This follows from item 2 with $x = b + 1$. We use b to be consistent with a later use.)

(5)

$$\frac{1}{d} + \frac{1}{f_a(d)} = \frac{1}{d-a+1}$$

$$+ \frac{1}{(d-a+1)(d-a)+a} + \frac{1}{f_a((d-a+1)(d-a)+a)}$$

(This follows from items 1 and 2.)

Proof. 1)

$$\frac{1}{d} + \frac{a}{d(d-a)} = \frac{d-a}{d(d-a)} + \frac{a}{d(d-a)} = \frac{d}{d(d-a)} = \frac{1}{d-a}$$

2) We use the following:

$$\frac{1}{f_a(x(x-1)+a)} = \frac{a}{(x(x-1)+a)(x(x-1))}$$

$$= \frac{1}{x(x-1)} - \frac{1}{x(x-1)+a}$$

Note that

$$\frac{1}{x} + \frac{1}{x(x-1)+a} + \frac{1}{f_a(x(x-1)+a)} =$$

$$\frac{1}{x} + \frac{1}{x(x-1)+a} + \frac{1}{x(x-1)} - \frac{1}{x(x-1)+a} =$$

$$\frac{1}{x} + \frac{1}{x(x-1)} = \frac{x-1}{x(x-1)} + \frac{1}{x(x-1)} = \frac{x}{x(x-1)} = \frac{1}{x-1}.$$

$$\square$$

Definition 13.4. Let $a \geq 1$, $n \geq 3$. A *nice (n,a)-sequence* is a nice n-sequence $(d_1, \ldots, d_n)$ such that:

(1) $d_{n-1} \equiv 0 \pmod{a}$
(2) $d_{n-2} < d_{n-1} - a + 1$
(3) $d_n = f_a(d_{n-1})$
(4) $d_{n-1} \geq a + 1$.

Lemma 13.3. *Let* $a \geq 1$, $n \geq 3$. *If there exists a nice n-sequence* $(b_1, \ldots, b_n)$ *such that* $b_n \equiv 0 \pmod{a}$ *and* $b_{n-1} \geq a+1$ *then there exists a nice* $(n+3, a)$-*sequence.*

Proof. Assume there exists a nice n-sequence $(b_1, \ldots, b_n)$ such that $b_n \equiv 0 \pmod{a}$. Let k be such that $b_n = ak$. Using this and Lemma 13.2.3: we have

$$\frac{1}{b_n} = \frac{1}{b_n + 1} + \frac{1}{b_n(b_n + 1) + a} = \frac{1}{b_n + 1} + \frac{1}{ak(b_n + 1) + a}$$

$$= \frac{1}{b_n + 1} + \frac{1}{a(k(b_n + 1) + 1)} + \frac{1}{f_a(a(k(b_n + 1) + 1))}$$

Hence

$$\frac{1}{b_1} + \cdots + \frac{1}{b_{n-1}} + \frac{1}{b_n + 1} + \frac{1}{a(k(b_n + 1) + 1)}$$

$$+ \frac{1}{f_a(a(k(b_n + 1) + 1))} = 1$$

Take $d_1 = b_1$, ..., $d_{n-1} = b_{n-1}$, $d_n = b_n + 1$, $d_{n+1} = a(k(b_n + 1) + 1)$, $d_{n+2} = f_a(a(k(b_n + 1) + 1))$.

Conditions 1,3,4 are clearly true. Condition 2 holds by easy algebra. $\square$

Theorem 13.2. *Let* $a \geq 1$, $n \geq 3$. *For all but a finite number of n, there exists a nice* (n, a)-*sequence.*

Proof. We prove this by induction on n. We do not know what the base case is; however, one can use our proof of the base case to find it.

Base Case: By Lemma 13.1.2 there exists $m \in \mathbb{N}$ and a nice m-sequence where the last term is composite. (In fact all terms are composite though we do no need that.) By Lemma 13.3 there exists a nice $(m+3, a)$-sequence.

Induction Step: Assume $(d_1, \ldots, d_{n-1}, d_n)$ is a nice (n, a)-sequence. Let $d_{n-1} = d$. Note that $d \equiv 0 \pmod{a}$ and $d_n = f(d)$. By Lemma 13.2.5:

$$\frac{1}{d_{n-2}} + \frac{1}{d_{n-1}} + \frac{1}{f_a(d_{n-1})} = \frac{1}{d_{n-2}} + \frac{1}{d_{n-1} - a + 1}$$

$$+ \frac{1}{(d_{n-1} - a + 1)(d_{n-1} - a) + a}$$

$$+ \frac{1}{f_a((d_{n-1} - a + 1)(d_{n-1} - a) + a)}$$

We claim that

$$(d_1, \ldots, d_{n-2}, d_{n-1} - a + 1, (d_{n-1} - a + 1)(d_{n-1} - a)$$

$$+ a, f_a((d_{n-1} - a + 1)(d_{n-1} - a) + a))$$

is a nice $(n + 1, a)$-sequence.

We first just prove that it's a nice n-sequence. Clearly the sum of the reciprocals adds to 1. Clearly $d_1 < \cdots < d_{n-2}$ inductively. We have $d_{n-2} < d_{n-1} - a + 1$ inductively since that is Condition 2 for nice (n, a)-sequences. By algebra

$$d_{n-1} - a + 1 < (d_{n-1} - a + 1)(d_{n-1} - a) + a$$

$$< f_a((d_{n-1} - a + 1)(d_{n-1} - a) + a))$$

We now prove the conditions for being a nice (n, a)-sequence. Since (the old) $d_{n-1} \equiv 0 \pmod{a}$, $d_{n-1} - a \equiv 0 \pmod{a}$ and hence

$$(d_{n-1} - a + 1)(d_{n-1} - a) + a \equiv 0 \pmod{a}.$$

We need $(d_{n-1} - a + 1) < ((d_{n-1} - a + 1)(d_{n-1} - a) + a) - a + 1$. This is true by algebra. We need $d_{n+1} = f(d_n)$. This is clearly true. $\square$

13.6   Improving the Base Case

Theorem 13.2 can be restated as follows:

Theorem 13.3. *There is a function $f(a)$ such that, for all a, for all $n \geq f(a)$, there exists an (n, a) sequence.*

The proof of Theorem 13.2 gives no hint as to what the function f looks like. I actually *liked* this aspect. It made for a very unusual base case since you didn't know where it started. Vince wanted an upper bound on f. We both thought that finding a bound would be difficult. As we were going over the proofs in the paper, a high school student, Dilhan Salgado, knocked on my door needing a project. On a lark we asked him to find a bound on $f(n)$. A week later he did! (Note that Dilhan won the Maryland Math Competition a few months later.) His proof is below.

Theorem 13.4. *Let $a \in \mathbb{N}$. For all $n \geq a^{O(a)}$ there exists a nice (n, a)-sequence.*

Proof. We need to find an $(a^{(a+o(1))a}, n)$-sequence for our base case. After that we use the induction step as in the proof of Theorem 13.2.

Claim 1: For all primes p there exists a nice sequence of length $\leq p^{O(p)}$ such that p divides the last term.

Proof of Claim 1: We define operations on nice sequences. These operations will do most of the work for us.

(1) Assume $(c_1, \ldots, c_n)$ and $(d_1, \ldots, d_m)$ are nice. We define

$$M(c_1, \ldots, c_n, d_1, \ldots, d_m) = (c_1, \ldots, c_{n-1}, c_n d_1, c_n d_2, \ldots, c_n d_m).$$

It is easy to see that the output of M is a nice $(n + m - 1)$-sequence.

(2) Assume $(c_1, \ldots, c_n)$ is n-nice. We define

$$E(c_1, \ldots, c_n) = (c_1, \ldots, c_{n-1}, c_n + 1, c_n(c_n + 1)).$$

It is easy to see that the output of E is a nice $(n + 1)$-sequence.

(3) Assume $\vec{c}$ is nice and ends with t. Assume p does not divide t. Let

$$\vec{d} = E(\vec{c})$$

$$\vec{c}_2 = M\left(\vec{c}, \vec{d}\right)$$

$$(\forall i \geq 3)[\vec{c}_i = M(\vec{c}_{i-1}, \vec{c})].$$

Let $F(\vec{c}) = \vec{c}_{p-1}$. It is easy to see that $F(\vec{c})$ is a nice $(p(n-1) - n + 3)$-sequence (we will later just bound this by pn). The last term of $F(\vec{c})$ is $t^{p-1}(t+1)$. Since p does not divide t, by Fermat's little theorem, the last term is $\equiv t+1 \pmod{p}$.

(4) Assume $\vec{c}$ is nice and ends with a number that is $\equiv t \pmod{p}$. Assume p does not divide t. Then $F^{(i)}(\vec{c})$ is a nice $\leq (pn)^i$-sequence whose last term is $\equiv t + i \pmod{p}$.

If $p = 2$ or $p = 3$ then we use the sequence $(2, 3, 6)$. Assume $p \geq 5$. Let $\vec{c} = (2, 3, 6)$. Let t be such that $6 \equiv t \pmod{p}$. Then $F^{(p-t)}(\vec{c})$ is a nice $(3p)^{p-t}$-sequence with last term $\equiv t+(p-t) \equiv 0 \pmod{p}$. Note that $(3p)^{p-t} \leq p^{O(p)}$.

End of Proof of Claim 1

Let $a = p_1^{e_1} \cdots p_L^{e_L}$. By Claim 1 we can create, for each $1 \leq i \leq L$, a nice $(p_i)^{O(p_i)}$-sequence.

$\vec{c}_i$ whose last term is divisible by p_i.

If $\vec{c}$ is a nice n_1-sequence with last term $\equiv 0 \pmod{p}$ and $\vec{d}$ is a nice n_2-sequence with last term $\equiv 0 \pmod{q}$ then $M(\vec{p}, \vec{q})$ is a nice $(n_1 + n_2 - 1)$-sequence with last term $\equiv 0 \pmod{pq}$. Hence

$$M(\vec{c}_1, \vec{c}_1, \ldots, \vec{c}_1, \vec{c}_2, \ldots, \vec{c}_2, \ldots, \vec{c}_n)$$

(where we take each c_i e_i times) is a nice sequence of length

$$\sum_{i=1}^{L} e_i(p_i)^{O(p_i)}$$

Since $e_i \leq \log a$, $L \leq \log a$, and $p_i \leq a$, this sum is

$$\leq (\log a)^2 a^{O(a)} \leq a^{O(a)}.$$

The last term is divisible by a. Since $a^{O(a)} + 3 \leq a^{O(a)}$ by Lemma 13.3, we have a nice $a^{O(a)}$-sequence. □

The proof of Theorem 13.4 gives the bound $n_0 \leq a^{O(a)}, a$. In Section 13.8 we give empirical evidence that indicates $n_0 \leq O(\log a)$.

13.7 Another Infinite Set of Proofs

I often discuss math with Darling in restaurants. Here is an excerpt from one such conversation:

Bill: Note that $\frac{1}{2} + \frac{1}{3} + \frac{1}{6} = 1$. Can you come up with a way to express 1 as the sum of 4 distinct reciprocals?

Darling: Yes! (She writes on a placement.) Just use $\frac{1}{6} = \frac{1}{8} + \frac{1}{24}$. OH, or I could use $\frac{1}{6} = \frac{1}{9} + \frac{1}{18}$. OH, or I could use $\frac{1}{6} = \frac{1}{10} + \frac{1}{15}$.

Bill: Oh! You just came up with three more proofs of the reciprocal theorem: Oh! — Only two more proofs. I will label the solution by what we use for the base case and by what we use for the induction step. For example (1) if we use $\frac{1}{2} + \frac{1}{3} + \frac{1}{6} = 1$ for the base case I denote that $(2, 3, 6)$, and (2) if we use $\frac{1}{6x} = \frac{1}{8x} + \frac{1}{24x}$ for the induction set I denote that $(6, 8, 24)$.
SOLUTION SIX(2,3,6)(6,8,24): We prove $(\forall n \geq 3)[P(n)]$. We load the induction hypothesis with the additional assumption that $d_n \equiv 0$ (mod 6). Use $(2,3,6)$ for the base case. We load the induction hypothesis with the additional assumption that $d_n \equiv 0$ (mod 6). Use

$$\frac{1}{6x} = \frac{1}{8x} + \frac{1}{24x},$$

to go from $P(n-1)$ to $P(n)$.

SOLUTION SIX(2,3,6)(6,9,18): Similar to SO-
LUTION SIX(2,3,6)(6,9,10).

SOLUTION SIX(6,10,150): Does not Work:
If we try to use $\frac{1}{6x} = \frac{1}{10x} + \frac{1}{15x}$. then we won't
get that the $d_n \equiv 0 \pmod 6$. So the equation $\frac{1}{6} = \frac{1}{10} + \frac{1}{15}$ does not lead to a new proof.

Darling: The two SOLUTION SIXs were based on $\frac{1}{6} = \frac{1}{8} + \frac{1}{24}$
and $\frac{1}{6} = \frac{1}{9} + \frac{1}{18}$.

Can you get other proofs by writing, for some large
d, $\frac{1}{d}$ as the sum of two reciprocals?

Bill: (Thinking) Yes! Here is an example.
SOLUTION SIX(2,3,5,8,56)(56,58,1624): We
prove $(\forall n \geq 5)[P(n)]$. Load the induction hypothesis with the additional assumption that $d_n \equiv 0$
$\pmod{56}$. We use the following for the base case:

$$\frac{1}{2} + \frac{1}{3} + \frac{1}{8} + \frac{1}{42} + \frac{1}{56} = 1$$

Use

$$\frac{1}{56x} = \frac{1}{58x} + \frac{1}{1624x}$$

(noting that 56 divides 1624) to go from $P(n-1)$
to $P(n)$.

Based on Darling's proof we give two proofs that a high school
student taking the exam could have given but just happened not
to, and then give an infinite number of proofs based on them.

Theorem 13.5. *For all $n \geq 3$, $P(n)$ holds.*

Proof. We use base3 for the base case. We load the induction hypothesis with the assumption that $d_n \equiv 0 \pmod 6$.

SOLUTION FIVE-a: Assume $(d_1, \ldots, d_n)$ is a nice sequence. Assume $d_n = 6d$. Then since $\frac{1}{6d} = \frac{1}{9d} + \frac{1}{18d}$ $(d_1, \ldots, d_{n-1}, 9d, 18d)$ is a nice sequence.

SOLUTION FIVE-b: Assume $(d_1, \ldots, d_n)$ is a nice sequence. Assume $d_n = 6d$. Then since $\frac{1}{6d} = \frac{1}{8d} + \frac{1}{24d}$ $(d_1, \ldots, d_{n-1}, 8d, 24d)$ is a nice sequence. $\qquad\square$

The next theorem generates an infinite number of proofs using the idea of Theorem 13.5.

Theorem 13.6.

(1) If there exists a nice sequence of length n_0 with composite last term then, for all $n \geq n_0$, $P(n)$.

(2) There exists an infinite number of nice sequences with composite last term.

Proof. 1) Let $(c_1, \ldots, c_{n_0})$ be a nice sequence with composite last term. Let e be a nontrivial factor of c_{n_0}. Note that:

- $\frac{1}{c_{n_0}} = \frac{1}{c_{n_0}(e+1)/e} + \frac{1}{c_{n_0}(e+1)}$ since e divides c_{n_0}, we are writing $\frac{1}{c_{n_0}}$ as the sum of two reciprocals, and
- $c_{n_0}(e+1)/e < c_{n_0}(e+1)$.

We prove that, for all $n \geq n_0$, there exists a nice sequence of length n with last term divisible by c_{n_0}.
Base Case: Use $(c_1, \ldots, c_{n_0})$.
Induction Step: Assume that there is a nice sequence of length n, $(d_1, \ldots, d_n)$ with $d_n \equiv 0 \pmod{c_{n_0}}$. Let $d_n = c_{n_0} x$. Then $\frac{1}{d_n} = \frac{1}{c_{n_0} x} = \frac{1}{c_{n_0} x(e+1)/e} + \frac{1}{c_{n_0} x(e+1)}$. Hence $(d_1, \ldots, d_{n-1}, c_{n_0} x(c_{n_0}+1)/e, c_{n_0} x(c_{n_0}+1))$ is a nice sequence of length $n+1$ with last term divisible by c_{n_0}.

2) This follows from Lemma 13.1.2. $\qquad\square$

In the proof of Theorem 13.6 we write $\frac{1}{c_{n_0}}$, with the aid of a divisor e, as $\frac{1}{x} + \frac{1}{y}$, where y divides c_{n_0}. In the table below we show what this sum looks like for $4 \le c_{n_0} \le 12$ and possible e.

c_{n_0}	e	$e+1$	$y = \frac{c_{n_0}(e+1)}{e}$	$\frac{1}{c_{n_0}} = \frac{1}{c_{n_0}(e+1)/e} + \frac{1}{c_{n_0}(e+1)}$
4	2	3	6	$\frac{1}{4} = \frac{1}{6} + \frac{1}{12}$
6	2	3	9	$\frac{1}{6} = \frac{1}{9} + \frac{1}{18}$
6	3	4	8	$\frac{1}{6} = \frac{1}{8} + \frac{1}{24}$
8	2	3	12	$\frac{1}{8} = \frac{1}{12} + \frac{1}{24}$
8	4	5	10	$\frac{1}{8} = \frac{1}{10} + \frac{1}{40}$
9	3	4	12	$\frac{1}{9} = \frac{1}{12} + \frac{1}{36}$
10	2	3	15	$\frac{1}{10} = \frac{1}{15} + \frac{1}{30}$
10	5	6	12	$\frac{1}{10} = \frac{1}{12} + \frac{1}{60}$
12	2	3	18	$\frac{1}{12} = \frac{1}{18} + \frac{1}{36}$
12	3	4	16	$\frac{1}{12} = \frac{1}{16} + \frac{1}{48}$
12	4	5	15	$\frac{1}{12} = \frac{1}{15} + \frac{1}{60}$
12	6	7	14	$\frac{1}{12} = \frac{1}{14} + \frac{1}{84}$

13.8   For $1 \le a \le 149$ What is the Smallest n_0?

In Theorems 13.2 and 13.4 we did, for each a, find an n_0 and a proof that for all $n \ge n_0$ there is a nice n-sequence. In Theorem 13.2 no bound on n_0 was given (though one could probably be derived), and in Theorem 13.4 we obtained $n_0 \le 2^a a^{(1+o(1))a}$.

We wrote a program that would, for $0 \le a \le 149$, find the first n_0 that works. Here is a summary of the data:

(1) For $a \in \{0, \ldots, 9\}$, $n_0 \in \{3, 4, 5\}$.
(2) For $a \in \{10, \ldots, 27\}$, $n_0 \in \{5, 6\}$.
(3) For $a \in \{28, \ldots, 149\}$, $n_0 \in \{6, 7\}$.

These numbers are far smaller than our current theoretical bounds of $a^{O(a)}$. While there is not much data to go on we conjecture that the actual growth rate is logarithmic.

13.9   Article vs. Blog

I may write this chapter up as an article for a Math Journal. If so it will be co-authored with Vince Cozzo, Darling Gasarch (I'll need to learn her first name), and Dilhan Salgado. When I first married Darling I told her that one of the goals of our marriage was to have a Cozzo-Gasarch-Gasarch-Salgado paper. And now we will!

For the article I will have to remove all of the description of how we got the result. I wonder how much math is lost to us because authors have to remove the process. Oh well.

13.10   The Students Answers to Part (a)

The students submitted 32 correct solutions to Part (a). We list all correct submitted solutions in lexicographic order, along with how many students submitted each one. We also note which of SOLUTION ONE, TWO, THREE, FOUR, FIVE-a, FIVE-b would lead to the answer they gave. For example, since we gave $(2, 3, 6)$ and $(2, 3, 7, 42)$ as solutions, and SOLUTION ONE takes $(2, 3, 7, 42)$ and produces $(2, 3, 7, 43, 1886)$, that solution to Part (a) is linked to SOLUTION ONE to Part (b).

Solution	Numb	Comment
(2,3,7,43,1806)	91	Linked to SOLUTION ONE.
(2,3,7,48,336)	3	
(2,3,7,56,168)	1	
(2,3,7,63,126)	6	Linked to SOLUTION THREE.
(2,3,7,70,105)	1	
(2,3,8,25,600)	1	
(2,3,8,30,120)	1	
(2,3,8,32,96)	6	Linked to SOLUTION FIVE-b
(2,3,8,36,72)	5	
(2,3,8,42,56)	11	
(2,3,9,21,126)	2	
(2,3,9,24,72)	4	
(2,3,9,27,54)	3	Linked to SOLUTION FIVE-a
(2,3,10,20,60)	5	
(2,3,11,22,33)	1	
(2,3,12,15,60)	1	
(2,3,12,16,48)	1	
(2,3,12,14,84)	2	Linked to SOLUTION FOUR.
(2,3,12,18,36)	12	Linked to SOLUTION TWO.
(2,4,5,25,100)	3	
(2,4,5,30,60)	1	
(2,4,6,14,84)	3	
(2,4,6,16,48)	1	
(2,4,6,18,36)	2	
(2,4,6,20,30)	1	
(2,4,7,12,42)	4	
(2,4,7,14,28)	2	
(2,4,8,12,24)	6	
(2,4,8,10,40)	2	
(2,5,6,10,30)	1	
(2,5,6,12,20)	2	
(3,4,5,6,20)	3	

Chapter 14

Is This Problem Too Hard for a High School Math Competition?

Prior Knowledge Needed: None.

14.1 The Point

How can you tell if a problem is too hard?

In the next section I state a problem that was rejected from the *University of Maryland High School Mathematics Competition* in 2011 because it was thought to be too hard. (For information on this competition see the Chapter 11.) I came up with it by knowing a lemma and working backwards to come up with a problem, so I couldn't tell how hard it was.

This gives rise to a question (solve the problem) and a meta question (is it too hard for a High School Competition?).

I present the problem and several solutions. The solutions are by Sam Zbarsky, David Eppstein, and I. Sam Zbarsky is a high school student, so he is a proof-by-example that a high school student could solve it. But (1) it took him half an hour and (2) he won the UMCP Math competition twice. That it took someone of his caliber half an hour might indicate that the problem is too hard. David Eppstein, professor of computer science, used specialized knowledge to solve the problem. My solution would be hard for a high school student to come up with since I knew a lemma that they likely would not.

After I posted it, some readers wrote a computer program to solve it. Note that a computer program would not be available to students in the math competition.

Was it too hard for the exam? I leave that as an open question.

Here is the problem:

> *Prove or disprove: there exist natural numbers*
> $n_1, \ldots, n_{10}$ *such that*
>
> - $n_1 + \cdots + n_{10} = 2011$, *and*
> - $\frac{1}{n_1} + \cdots + \frac{1}{n_{10}} = 1$.

We will need the following definition.

Definition 14.1. Let $n \in \mathbb{N}$. $cool(n)$ is the least m such that there exists $n_1, \ldots, n_m \in \mathbb{N}$ with

- $n = n_1 + \cdots + n_m$
- $1 = \frac{1}{n_1} + \cdots + \frac{1}{n_m}$.

Note 14.1. $cool(n) \le n$ by writing n as the sum of m 1's.

14.2 Solution by William Gasarch

I happened to come across the following lemma. I don't recall when or how.

Lemma 14.1.

(1) $cool(20n + 11) \le cool(n) + 3$.
(2) $cool(2n + 8) \le cool(n) + 2$.
(3) $cool(4n + 6) \le cool(n) + 2$.

Proof. We assume $cool(n) = m$. Let $n_1, \ldots, n_m$ be such that

$$n = \sum_{i=1}^{m} n_i \qquad \sum_{i=1}^{m} \frac{1}{n_i} = 1.$$

1) $20n + 11 = 2 + 4 + 5 + \sum_{i=1}^{m} 20n_i$

$\frac{1}{2} + \frac{1}{4} + \frac{1}{5} + \sum_{i=1}^{m} \frac{1}{20n_i} = \frac{19}{20} + \frac{1}{20} = 1.$

2) $2n + 8 = 4 + 4 + \sum_{i=1}^{m} 2n_i$

$\frac{1}{4} + \frac{1}{4} + \sum_{i=1}^{m} \frac{1}{2n_i} = \frac{1}{2} + \frac{1}{2} = 1.$

3) $4n + 6 = 2 + 4 + \sum_{i=1}^{m} 4n_i$

$\frac{1}{2} + \frac{1}{4} + \sum_{i=1}^{m} \frac{1}{4n_i} = \frac{3}{4} + \frac{1}{4} = 1$ □

We now use the lemma to show that $\text{cool}(2011) \leq 10$. Since $2011 = 20 \times 100 + 11$,

$$\text{cool}(2011) \leq \text{cool}(100) + 3.$$

Since $100 = 2 \times 46 + 8$,

$$\text{cool}(100) \leq \text{cool}(46) + 2.$$

Since $46 = 4 \times 10 + 6$,

$$\text{cool}(46) \leq \text{cool}(10) + 2.$$

Since $10 = 4 + 4 + 2$ and $\frac{1}{4} + \frac{1}{4} + \frac{1}{2} = 1$,

$$\text{cool}(10) = 3.$$

Building everything back up we have:

$$\text{cool}(46) \leq \text{cool}(10) + 2 = 5$$

$$\text{cool}(100) \leq \text{cool}(46) + 2 = 5 + 2 = 7$$

$$\text{cool}(2011) \leq \text{cool}(100) + 3 = 7 + 3 = 10$$

What are $n_1, \ldots, n_{10}$**?** The proof is entirely constructive so we use it to find the numbers, though this is not required for the solution. Since $10 = 4 + 4 + 2$

$$\begin{aligned}
46 &= 2 + 4 + 4 \times 10 \\
&= 2 + 4 + 4(4 + 4 + 2) \\
&= 2 + 4 + 16 + 16 + 8 \\
&= 16 + 16 + 8 + 4 + 2
\end{aligned}$$

$$100 = 4 + 4 + 2 \times 46$$
$$= 4 + 4 + 2(16 + 16 + 8 + 4 + 2)$$
$$= 4 + 4 + 32 + 32 + 16 + 8 + 4$$
$$= 32 + 32 + 16 + 8 + 4 + 4 + 4$$

$$2011 = 2 + 4 + 5 + 20 \times 100$$
$$= 2 + 4 + 5 + 20(32 + 32 + 16 + 8 + 4 + 4 + 4)$$
$$= 2 + 4 + 5 + 640 + 640 + 320 + 160 + 80 + 80 + 80$$
$$= 640 + 640 + 320 + 160 + 80 + 80 + 80 + 5 + 4 + 2$$

One can check that the sum of the reciprocals is 1.

14.3 Solution by Sam Zbarsky

Sam emailed me the following proof, hence the presentation is in his voice. Although his solution is short, it took him half an hour to find.

Proof. I noticed that if I include $2, 4, 5$ then I need 7 numbers that sum to 2000 whose reciprocals sum to $\frac{1}{20}$. If all are divisible by 20 then the problem is

$$20y_1 + \cdots + 20y_7 = 2000$$

$$\frac{1}{20y_1} + \cdots + \frac{1}{20y_7} = \frac{1}{20}$$

which can be expressed as

$$y_1 + \cdots + y_7 = 100$$

$$\frac{1}{y_1} + \cdots + \frac{1}{y_7} = 1$$

Then I guessed and checked until I got $2, 6, 8, 15, 15, 24, 30$. Hence the final solution is: $2, 4, 5, 40, 120, 160, 300, 300, 480, 600$. □

14.4   Solution by David Eppstein

David Eppstein emailed me the following proof, hence the presentation is in his voice.

Proof. I used Sylvester's sequence[1]

$$\frac{1}{2} + \frac{1}{3} + \frac{1}{7} + \frac{1}{43} + \frac{1}{1806} = 1$$

$$2 + 3 + 7 + 43 + 1806 = 1861$$

I tried to modify it to make the number of terms larger and the difference from 2011 smaller. These five terms have sum 1861, which is off by 150. I noticed that replacing $\frac{1}{2}$ by $\frac{1}{4} + \frac{1}{4}$ makes the difference an even rounder number, 144. So now I have

$$\frac{1}{3} + \frac{1}{4} + \frac{1}{4} + \frac{1}{7} + \frac{1}{43} + \frac{1}{1806} = 1$$

$$3 + 4 + 4 + 7 + 43 + 1806 = 1867$$

I observed (along with several similar observations) that if I could find five integers whose reciprocals summed to 1 and whose sum was 37, then I could use that solution to replace one of the $\frac{1}{4}$ terms to get a solution to the whole problem. (We'll see how later.)

I then built a big table of all the sets of four unit fractions summing to one that I could think of, and computed the difference of the sum of denominators of each with 37. When I found that

$$2 + 4 + 5 + 20 = 31$$

which differs from 37 by 6

$$\frac{1}{2} + \frac{1}{4} + \frac{1}{5} + \frac{1}{20} = 1$$

---

[1] *Sylvester's Sequence* is an integer sequence in which each number is the product of all prior terms, plus 1.

I realized that I could substitute $\frac{1}{2} = \frac{1}{4} + \frac{1}{4}$ to get

$$4 + 4 + 4 + 5 + 20 = 37$$

$$\frac{1}{4} + \frac{1}{4} + \frac{1}{4} + \frac{1}{5} + \frac{1}{20} = 1$$

Multiply all of the numbers by 4 to get

$$16 + 16 + 16 + 20 + 80 = 148$$

$$\frac{1}{16} + \frac{1}{16} + \frac{1}{16} + \frac{1}{20} + \frac{1}{80} = \frac{1}{4}$$

In

$$\frac{1}{3} + \frac{1}{4} + \frac{1}{4} + \frac{1}{7} + \frac{1}{43} + \frac{1}{1806} = 1$$

$$3 + 4 + 4 + 7 + 43 + 1806 = 1867$$

replace 4 by 16,16,16,20,80 to get

$$\frac{1}{3} + \frac{1}{4} + \frac{1}{7} + \frac{1}{16} + \frac{1}{16} + \frac{1}{16} + \frac{1}{20} + \frac{1}{43} + \frac{1}{80} + \frac{1}{1806} = 1$$

$$3 + 4 + 7 + 16 + 16 + 16 + 20 + 43 + 80 + 1806 = 2011$$

$\square$

Would a High School Student know Sylvester's sequence? I doubt it. However, they may recall Eppstein starting point,

$$\frac{1}{2} + \frac{1}{3} + \frac{1}{7} + \frac{1}{43} + \frac{1}{1806} = 1$$

was in the 2010 UMCP High School Math Competition, just the year before. The question there, was to find five numbers whose reciprocals add to 1.

14.5 More Bounds on cool(n)

Anonymous, who posts a lot, posted the following:

Lemma 14.2.

(1) cool($2n + 9$) $\leq$ cool(n) $+ 2$.
(2) cool($3n + 6$) $\leq$ cool(n) $+ 2$.
(3) cool($3n + 8$) $\leq$ cool(n) $+ 2$.
(4) cool($2n + 8$) $\leq$ cool(n) $+ 2$. *(Proven in Lemma 14.1.)*
(5) cool($4n + 6$) $\leq$ cool(n) $+ 2$. *(Proven in Lemma 14.1.)*
(6) cool($20n + 11$) $\leq$ cool(n) $+ 3$. *(Proven in Lemma 14.1.)*

Proof. We assume cool(x) $= n$. Let $n_1, \ldots, n_m$ be such that

$$n = \sum_{i=1}^{m} n_i \qquad \sum_{i=1}^{m} \frac{1}{n_i} = 1.$$

1) $2n + 9 = 9 + \sum_{i=1}^{m} 2n_i = 3 + 6 + \sum_{i=1}^{m} 2n_i$.
$\frac{1}{3} + \frac{1}{6} + \sum_{i=1}^{m} \frac{1}{2n_i} = \frac{1}{2} + \frac{1}{2} = 1$.

2) $3n + 6 = 6 + \sum_{i=1}^{m} 3n_i = 3 + 3 + \sum_{i=1}^{m} 3n_i$.
$\frac{1}{3} + \frac{1}{3} + \sum_{i=1}^{m} \frac{1}{3n_i} = \frac{2}{3} + \frac{1}{3} = 1$.

3) $3n + 8 = 8 + \sum_{i=1}^{m} 3n_i = 2 + 6 + \sum_{i=1}^{m} 3n_i$.
$\frac{1}{2} + \frac{1}{6} + \sum_{i=1}^{m} \frac{1}{3n_i} = \frac{2}{3} + \frac{1}{3} = 1$. $\qquad\square$

14.6 Recap

Here are all the solutions from readers. We present them in lexicographic order.

- Sam Zbarsky: 2,4,5,40,120,160,300,300,480,600
- Anonymous: 2,4,5,50,100,100,250,500,500,500 (by program)
- William Gasarch: 2,4,5,89,89, 89, 160,320,640,640
- David Eppstein: 3,4,7,16,16,16,20,43,80,1806
- Anonymous: 5,5,6,8,8,12,15,32,960,960 (by program)

- Anonymous: 6,6,10,10,12,15,15,15,62,1860 (by program)
- Anonymous: 6,8,8,8,12,15,16,16,62,1860 (by program)

A reader, named Lew, found, using a computer program, that there were 215,067 solutions to

$$n_1 + \cdots + n_{10} = 2011 \qquad \frac{1}{n_1} + \cdots + \frac{1}{n_{10}} = 1.$$

We do not list them in the appendix.

14.7 Open Questions

(1) How does $\mathrm{cool}(n)$ grow asymptotically? We suspect $\mathrm{cool}(n) \ll n$ but cannot even prove that $\mathrm{cool}(n) \le n - 1$.

(2) Consider the equations:

$$n_1 + \cdots + n_m = n \qquad \frac{1}{n_1} + \cdots + \frac{1}{n_m} = 1.$$

How many solutions are there? How many solutions are there asymptotically in n?

(3) Is the following question too hard for a high school math competition?

Prove or disprove: there exist distinct natural numbers $n_1, \ldots, n_{10}$ such that

- $2011 = n_1 + \cdots + n_{10}$, and
- $1 = \frac{1}{n_1} + \cdots + \frac{1}{n_{10}}$.

These questions could be asked with the stipulation that the n_i are distinct. As a starting point, Ronald Graham [Graham (1963)] has shown that, for all $n \ge 78$, there exists m and *distinct* $n_1, \ldots, n_m$ such that $n_1 + \cdots + n_m = n$ and $\frac{1}{n_1} + \cdots + \frac{1}{n_m} = 1$. The statement is false for $n = 77$. His proof did not give bounds on m; however, they would not be hard to derive from his proof. His base case is $n = 78, 79, \ldots, 333$. Hence, unless there is different proof, a question about sums of distinct numbers whose reciprocals add up to 1 is not appropriate for the UMCP Math Competition.

References

Graham, R. (1963). A theorem on partitions, *Journal of the Australian Math Society* **3**, http://www.math.ucsd.edu/~ronspubs/.

Chapter 15

What is an Elegant Proof?
A Theorem About Digits

Prior Knowledge Needed: None.

15.1 Point

Paul Erdős thought that there was a book where God had written down the most elegant proofs for mathematical theorems. What is an elegant proof? I doubt that elegance can be defined rigorously; however, mathematicians have a sense of when a proof is elegant. Consider the following theorem, which we refer to as *the main theorem.*

The only numbers that are the sum of the squares of their digits are 0 and 1.

One measure of elegance is the number of cases in a proof. We give three proofs of the 2-digit case and discuss their elegance. We then prove the main theorem.

15.2 The Two-Digit Lemma

Lemma 15.1. *No 2-digit number is the sum of the squares of its digits.*

Proof One: Exhaustive Search

We wrote a program that checked *all* numbers between 10 and 99. None of them were the sum of the squares of its digits.
End of Proof One

Number of cases: 90. This proof is not elegant by our measure. Nor would most people consider it elegant. The proof is hard to verify. You cannot give it to someone else to check. You could give them the program to check. Verifying programs is considered a hard problem; however, this program is short enough that it probably can be verified. I wonder if the verification proof is elegant.

Proof Two: Factorization Assume, by way of contradiction, that $1 \leq a \leq 9$, $0 \leq b \leq 9$, and $10a + b = a^2 + b^2$. We rewrite this as $b(b - 1) = a(10 - a)$.
Case 1: If a is odd then $a(10 - a)$ is odd. For all b, $b(b - 1)$ is even. Hence a cannot be odd.
Case 2: $a = 2, 8$. Then $b(b - 1) = 16$. There are no factorizations of 16 where the factors differ by 1.
Case 3: $a = 4, 6$. Then $b(b - 1) = 24$. There are no factorizations of 24 where the factors differ by 1.
End of Proof Two
Number of cases: 3. The proof has one theme, factorization, that is used repeatedly. This has a certain elegance. In addition, one feels that some actual mathematics is being done.

Proof Three: Mod 2, Mod 4, and Mod 10
Assume, by way of contradiction, that there exists $1 \leq a \leq 9$ and $0 \leq b \leq 9$ such that

$$10a + b = a^2 + b^2 \text{ henceforth } \textit{the main equation.}$$

Take this mod 2 using $x^2 \equiv x \pmod 2$ and $10 \equiv 0 \pmod 2$ to obtain

$$a \equiv 0 \pmod 2.$$

If we take the main equation mod 4, using $a^2 \equiv 0 \pmod 4$, we get

$$b(b - 1) \equiv 0 \pmod 4.$$

Hence $b \in \{0, 1, 4, 5, 8, 9\}$.

We now take the main equation mod 10.

$$b \equiv a^2 + b^2 \quad (\text{mod } 10).$$

$$a = \sqrt{-b(b-1)} \quad (\text{mod } 10).$$

We now run through $b \in \{0, 1, 4, 5, 8, 9\}$ and note for which ones $a = \sqrt{-b(b-1)}$ (mod 10) exists and is a nonzero even number. Note that the squares mod 10 are $0, 1, 4, 5, 6, 9$ so the only nonzero even ones are $4, 6$.

b	$-b(b-1)$ (mod 10)	$a = \sqrt{-b(b-1)}$ (mod 10)	comment
0	0	0	NO GOOD: Need $1 \leq 1 \leq 9$
1	0	0	NO GOOD: Need $1 \leq 1 \leq 9$
4	8	NONE	NO GOOD: No Square Root
5	0	0	NO GOOD: Need $1 \leq 1 \leq 9$
6	0	0	NO GOOD: Need $1 \leq 1 \leq 9$
8	4	2, 8	OH, Will need to check
9	8	NONE	NO GOOD: No Square Root

So the only candidates for (a, b) are (2,4), (8,4). One can check that these do not satisfy the main equation.

End of Proof Three

Number of cases (number of rows in the table): 7. Similar to proof two, there is one theme and one feels that some actual mathematics is being done.

Proofs two and three are far more elegant than proof one, either by our measure or by common sense. Also note that proofs two and three give an intuition for why the theorem is true, whereas proof one does not.

Which proof is more elegant, proof two or proof three? We say proof two since there are fewer cases *and* some of the cases in proof three are long.

15.3   The Main Theorem

Theorem 15.1. *The only numbers that are the sum of the squares of their digits are 0 and 1.*

Proof. Let x be a number.
Case 1: x is a 1-digit number. If $x \geq 2$ then $x < x^2$. Hence x cannot be the sum of the squares of its digits.
Case 2: x is a 2-digit number. Then x is not the sum of the squares of its digits by Lemma 15.1.
Case 3: x is a 3-digit number. Assume

$$x = 100a + 10b + c = a^2 + b^2 + c^2, \text{ where } 1 \leq a \leq 9, 0 \leq b, c \leq 9.$$

The largest x can be is $9^2 + 9^2 + 9^2 < 300$. Hence $a \in \{1, 2\}$.
So, the largest x can be is $2^2 + 9^2 + 9^2 = 176 < 200$. Hence $a = 1$ and $b \leq 7$.
So, the largest x can be is $1^2 + 7^2 + 9^2 = 141$. Hence $b \leq 4$.
So, the largest x can be is $1^2 + 4^2 + 9^2 = 98 < 100$. Hence there is no such x.
Case 4: x has $m \geq 4$ digits. Let $x = d_{m-1} \cdots d_0$ in base 10. Since $d_{m-1} \geq 1$ we have $x \geq 10^{m-1}$. Since x is the sum of the squares of its digits, $x \leq 81m$. Summing up

$$10^m \leq x \leq 81m.$$

We leave it to the reader to show that if $m \geq 4$ then this inequality cannot hold.   □

There are four cases; however, Case 2 uses a Lemma that has five cases. So we'll say
Number of cases: 9.

15.4   For the Reader

We leave it to the reader to look at when numbers can be the sum of the pth powers of their digits in base b. May all your proofs be elegant even if we cannot define elegance rigorously.

Chapter 16

Is This Problem Interesting?

Prior Knowledge Needed: None.

16.1 The Point

Nate Silver is a pollster who runs a website [Silver (2008–)] which is mostly about politics and sports (sometimes it's hard to tell the difference). He also has on it, a problems column called *The Riddler*, edited by Oliver Roeder. I sometimes look at and try to do the problems. They fall into several categories:

(1) Problems I have seen before, perhaps disguised.
(2) Problems that I have not seen before but are so clearly in my wheelhouse (e.g., Ramsey Theory, though disguised) that I could solve them.
(3) Problems that look interesting that I don't know how to solve and are worth working on since the answer is interesting.
(4) Problems that look interesting that I don't know how to solve, but are not worth working on since the answer is not interesting.

The issue with these problems is that I can't tell Type 3 from Type 4. So I may be interested in a problem, work on it, not solve it (which is fine), but then look at the answer and not be enlightened.

I will present a problem that is either Type 3 or 4 (could be a matter of taste) and trace through my thoughts — both on how to solve it and on if it's interesting.

16.2 The Problems

In the July 28, 2017 Riddler Column [Roeder (2017b)] the following question was posed (I paraphrase):

Definition 16.1. Let $A \subseteq \mathbb{N}$. A sequence is A-nice (just nice if A is understood) if

(1) Every element in the sequence is in A.
(2) No element appears more than once in the sequence.
(3) Every element is either a factor or multiple of the previous element (except the first element which has no previous element).

Example: 17, 1, 47, 94, 2, 4, 20, 5, 10, 90, 3, 21, 7, 42, 6, 30, 15

Problem: Let $A = \{1, \ldots, 100\}$. Find the longest A-nice sequence.

In case you want to work on it yourself my answer is on the next page (it is not nearly optimal). Warning: you probably cannot get the optimal solution or anywhere near it without a computer program. Even with one it may be difficult.

16.3 My Solution and Interest Level

After much work and doodling I found a nice sequence of length 42. I present it as a sequence of sequences that are either increasing or decreasing:

88 44 22 11
33 66
6 84
7 14 70
35 5
15 30 90
10 60
20 40 80
2 28
4 56
8 16 32 96
3 9 18 36
12 72
24 48
16 1
13 26 52

Was this interesting?

YES — I personally enjoyed working on it.
NO — It was *ad hoc*. There is no way I could formalize what I did for a $\{1, \ldots, n\}$-nice sequence.

Is 42 the best you can do? Fans of *The Hitchhiker's Guide to the Galaxy* would say *of course!*. Alas, it is not the answer. The next section reveals what *The Riddler* said.

16.4 The Riddler Said ... Was it Interesting?

In the August 4 2017 Riddler Column [Roeder (2017a)] the answer was revealed! There is a nice sequence of length 77:

93 31 62 1

87 29
58 2
92 46 23
69 3
57 19
38 76
4 68
34 17
85 5
35 70
10 100
50 25
75 15
45 90
30 60
20 40 80
16 64
32 96
48 24 12 6
78 26
52 13
91 7
49 98
14 56
28 84
42 21
63 9
81 27
54 18
36 72
8 88
44 22
66 33

99 11

55

They also showed another sequence of length 77. How were they found? Are they optimal? They were found by a computer program. The submitters claimed that the program proved it was optimal. Since two different people got it, we are inclined to believe them; however, that is not a proof.

Is this interesting?

YES — we know the answer.

NO — we do not have a technique for general n.

The problem they discuss is as follows: Given n, find the largest $\{1, \ldots, n\}$-nice sequence. This problem can be restated as finding the longest path in a particular graph. The column brings up the issue of NP-completeness. However, this does not apply for two conflicting reasons:

(1) The problem of given a *general* graph find its longest path is indeed NP-complete. However, they are asking this for a particular type of graph. For this type of graph the problem might be in P. There might even be a simple formula, e.g, $7(\lceil \sqrt{n} + 1 \rceil)$.

(2) NP-completeness is about the time of an algorithm as a function of the input length. For the problem at hand the input is n which has length $\lg n$. The graph produced is exponential in the size of the input. Hence the problem is *not* obviously in NP. The problem may be much harder than NP.

16.5 What Makes a Good Puzzle

What makes a good puzzle? There is no single answer; however, it should have at least four of the following six points.

(1) The puzzle should be easy to state.

(2) The puzzle should have a clever solution.

(3) Neither the puzzle nor the answer should require techniques beyond the person solving the puzzle. A good puzzle for a Math PhD might be a bad puzzle for a kindergarten student, unless he's Terry Tao.

(4) If someone cannot solve the puzzle then they should be able to understand and appreciate the solution. Martin Gardner called this *the AHA moment*. Archimedes called this *the Eureka moment*.

(5) If a problem requires use of a computer, or recursive algebraic topology, the solver should be warned.

(6) If a problem is a trick question the solver should be warned. (For some examples of trick questions see Chapter 18.)

16.6   Research Questions Inspired by the Puzzle

What is the complexity of the following problems.

(1) On input n, find the length of the longest $\{1, \ldots, n\}$-sequence.

(2) On input n, find a longest $\{1, \ldots, n\}$-sequence.

(3) On input n, find how many longest $\{1, \ldots, n\}$-sequences there are.

(4) On input set A, find the length of the longest A-sequence.

(5) On input set A, find a longest A-sequence.

(6) On input set A, find how many longest A-sequences there are.

As far as I know these are all open. Are they interesting? I do not know. I may be making the same mistake that *The Riddler makes*: posting a problem whose answer is not interesting.

References

Roeder, O. (2017a). Is this bathroom occupied, *Fivethirtyeight.com* `https://fivethirtyeight.com/features/is-this-bathroom-occupied/`.

Roeder, O. (2017b). Pick a number any number, *Fivethirtyeight.com* `https://fivethirtyeight.com/features/pick-a-number-any-number/`.

Silver, N. (2008–). Fivethirtyeight.com, `https://fivethirtyeight.com/`.

Chapter 17

A Hat Problem, its Generalizations, and a Story!

Prior Knowledge Needed: For Sections 17.4, 17.5 it would be helpful, but not necessary, to know partitions and equivalence classes.

A Hat Problem: Assume n people are standing in a line, each facing the person in front of him. Each person is wearing either a RED hat or a BLUE hat. We will call the person who sees everyone in front of him the first person. He sees all of the hats except his own. The next person, who sees all but two hats, we call the second person, etc. The last person cannot see any hats.

One at a time, starting from the first person, each person shouts out a hat color (either RED or BLUE). The goal for the entire group is to maximize how many people shout their own hat color. They CAN meet ahead of time to discuss a strategy.

Mr. Bad's goal is to is to minimize how many people shout their own hat color. Mr. Bad, who knows the strategy the people will use (by listening in on their conversation), places the hats on people to minimize how many they get right.

Here is a strategy: The first person shouts RED if at least half the hats he sees are RED, and BLUE otherwise. Then everyone, in turn, shouts that color. They are guaranteed to get (at least) $\lfloor n/2 \rfloor$ correct.

Here is another strategy: The first person shouts the color of second person, who then shouts that color, and gets it right;

the third person shouts the color of the fourth person, who then shouts that color, and gets it right; etc. They are guaranteed to get (at least) $\lfloor n/2 \rfloor$ correct.

Can they do better than $n/2$? Can one prove that they can't beat $n/2$? Perhaps the answer is something strange, like they can do $n - \lceil \lg n \rceil$ but no better than that.

Generalizations:

c **Colors:** This is the same as the original hat problem except that there can be any number $c \geq 2$ colors of hats. You can assume that the colors are known in advance. For clarity we will refer to the original problem as the 2-color hat problem.

Infinite Colors: This is the same as the original 2-color hat problem except that there are now an infinite number of colors of hats. You can assume that the colors are known in advance.

The inclusion of the *Infinite Colors* problem is an accident. An earlier draft of this chapter did not explain the next problem well and the reviewer thought it was what is now the *Infinite Colors* problem. Hence, I got to thinking about the *Infinite Colors Problem* and I now include it. Some ideas can happen by accident!

c **Colors, Infinite People:** This is the same as the c-color hat problem except that there are an infinite number of people. The first person in line sees the hats of the second, third, etc. The second sees the hats of the third, fourth, etc.

They can easily get an infinite number of hats right using an earlier strategy: The first person shouts the hat color of the second person, who then shouts that color (and gets it right); the third person shouts the hat color of the fourth person, who then shouts that color (and gets it right); etc. Alternatively, the first person can shout a color that appears infinitely often and then everyone else shouts that color. Can they get all but a finite number of hats right? Can they get all but some large number right? Can they get all but one right?

Infinite Colors, Infinite People: This is the same as the previous problem, except that there are also an infinite number of colors. They can easily get an infinite number of hats right using the first technique described in the previous problem. Can they get all but a finite number of hats right?

We put in a page break so that the reader can attempt these problems without seeing the solutions.

17.1   Three Solutions to the 2-Color Hat Problem

In my experience giving out this problem, people commonly find
one of the solutions in the introduction that gets at least $\lfloor n/2 \rfloor$
hats right. Here are three better solutions that people often find
after further thought. The last one is optimal.

Solution One: Assume 3 divides n. Partition the people into
contiguous groups of size three. Within each group do the following: The first person shouts RED if the next two people have
the same hat color, and BLUE if they have different hat colors.
The second person can now deduce his own hat color and shout
it out. The third person knows whether his hat is the same
color as the second person's, and now knows the hat color of the
second person's so he can deduce his own hat color and shout it
out. In this way, $\lfloor 2n/3 \rfloor$ people get the color of their hat correct.
End of Solution One

Solution Two: Since there are at most n RED hats, the number of RED hats can be represented with just $\lceil \lg(n) \rceil$ bits. The
first $\lceil \lg(n) \rceil$ people communicate the number of RED hats within
the first $n - \lceil \lg(n) \rceil$ people (and therefore also the number of
BLUE hats) by using RED and BLUE to represent 1 and 0.
Then each of the remaining $n - \lceil \lg(n) \rceil$ people can deduce the
color of his own hat by how many hats of his color he sees in
front of him and how many hats of his color he hears behind
him. Around $n - \lg(n)$ people will get their correct hat colors.
This can be refined to about $n - \lg(n - \lg n))$.
End of Solution Two

Solution Three: The first person counts how many RED hats
he sees. If it is even he shouts RED and if it is odd he shouts
BLUE. Each subsequent person can deduce his own hat color by
knowing just the parity of how many RED hats he sees in front
of him and how many RED hats he hears behind him.

Then $n - 1$ people get their correct hat color. This is clearly optimal since for any scheme Mr. Bad can make sure that the first person gets his own hat color wrong.

End of Solution Three

17.2   A Point Related to the 2-Color Hat Problem

I once had a group of students in a summer program working on the 2-color hat problem. The plan was that they would first do this problem and then move on to other hat problems. They solved the problem for $n = 3, 4, 5, 6$. For each of these they had a table (note the word *table* for later) of progressively larger cases. The $n = 6$ write up seemed to be Solution Two above, though written out as a table without really seeing what was going on.

I let them go on (perhaps too long) since they kept telling me *No Bill, Don't tell us how to do it!!* I was hoping they would have a breakthrough. But by the end of the second week they still hadn't gotten it. These were good students but they were "down a rabbit hole."

I finally told them Solution Three above. I *thought* they would say *Oh, that's very nice, now lets see if we can do something similar for c colors.* But no. They insisted that their solution using tables was *more intuitive and more informative.* None of that is remotely true. What is true is that by that point they were emotionally invested in their tables.

All summer I told them *DUMP YOUR TABLES* now that you have a better way of doing it. They never did. Their final talk still presented their solution via tables, and Solution Three above, as a different but equally interesting solution.

I now use the phrase *DUMP YOUR TABLES* to mean *DUMP SOME OLD WAY OF DOING THINGS THAT YOU ARE EMOTIONALLY ATTACHED TO BUT REALLY DOES NOT WORK.* Once you are aware of this phenomenon you will see it often.

You have a proof that uses a certain technique that you like (in my case, perhaps Ramsey Theory) but then a better proof comes along. You have to admit that the new proof is better. *DUMP YOUR TABLES.*

You get emotionally attached to a certain way to teach a course. Times change, technology changes; perhaps you should *DUMP YOUR TABLES.*

I have an idea for a blog entry that seems really good. I begin writing it, but it just isn't working. I *DUMP MY TABLES.* This happens often.

See the last section of this chapter for a personal example of *DUMPING MY TABLES.*

17.3 One Solution to the *c*-Color Hat Problem and Another Point

Solution: Number the hat colors $0, 1, 2, \ldots, c - 1$. The first person computes the sum mod c of all of the other hats, and shouts out the associated color. Since the second person knows both the sum of hat colors in front of him and the color that the first person shouted, he can deduce his own hat color. It is easy to see that the ith person, having heard the $i - 1$ people behind him and seeing the hats in front of him, has enough information to deduce his own hat color in the same way. Once again, they will get $n - 1$ correct, which is clearly optimal.
End of Solution:

I have given the problem to many people (and groups of people) who have seen and understood the 2-color problem, but were unable to solve the 3-color problem. The difficulty might be that we don't recognize the 2-hat problem in terms of mod-2, but instead in terms of *even* and *odd*. Hence going to mod-3 is a quantum leap.

17.4 Two Solutions to the Infinite Colors Hat Problem and Still Another Point

Number the hat colors $0, 1, 2, \ldots$, (which you can do since you know the colors in advance).

Solution One: The first person computes the sum of the hat colors of everyone in front of him. Note that he does not need to do the sum in modular arithmetic since all numbers are possible colors. It is easy to see that the ith person, having heard he $i - 1$ people behind him and seeing the hats in front of him, has enough information to deduce his hat color in the same way. Once again, they will get $n - 1$ correct, which is clearly optimal.
End of Solution One

Solution Two: Ahead of time the people agree on a bijection f between $\mathbb{N}^{n-1}$ and $\mathbb{N}$. The first person computes f on the tuple of hat colors of the remaining $n - 1$ people and shouts out that number. Now everyone knows his hat color.
End of Solution Two

Which solution is better? This is a matter of taste. The first one uses smaller hats (formally smaller hat numbers). It is also more elementary and requires less memorization. The second one is nice in that after the first person says his hats everyone else knows their hat color and does not need to listen to anymore hat colors.

17.5 Two Solutions to the c-color, Infinite People Hat Problem

Number the hat colors $0, 1, 2, \ldots, c - 1$
Solution One: In this solution all but a finite number of people will get their hat color correct.

Let R be the following relation on infinite sequences of hat colors: $R(\alpha, \beta)$ is TRUE iff α and β differ in only a finite number of places. This is an equivalence relation, and hence induces a partition. Every person knows this and they all agree on a representative from each partition (this requires the Axiom of Choice).

The hats are put on. Each person does the following: He sees the infinite number of hats in front of him, and is missing only a finite number of hats behind him. Hence he knows in which equivalence class the sequence of hats is. He remembers the representative from that class. He shouts the hat color he has in that representation.

They are all correct in their assessment of which partition the sequence is in, so they all use the same representative. Hence the sequence they produce differs only finitely from the actual one. Hence their sequence of shouts will be wrong only finitely often.

End of Solution One

Is there a solution where the number of people who get their hat wrong is bounded by some (perhaps large) constant? Is there a solution where the number of people who get their hat wrong is at most one? When I first saw this problem I assumed one could do no better than an unbounded but finite number of people getting their hat color wrong. But I was wrong!

Solution Two: In this solution only one person gets his hat color wrong.

Let R be as in Solution One. As in Solution One everyone knows which equivalence class the sequence of hats is in. Let the representative of that sequence be

$$\vec{a} = a_1 a_2 a_3 \cdots$$

Let the actual sequence of hats be

$$\vec{b} = b_1 b_2 b_3 \cdots$$

The first person yells

$$d \equiv \sum_{i=2}^{\infty} a_i - b_i \pmod{c}.$$

Since $\vec{a}$ and $\vec{b}$ differ finitely often, d is well defined.

We show how the second person can deduce his hat color. The second person computes

$$d' \equiv \sum_{i \geq 3} a_i - b_i \pmod{c}$$

and then yells out b_2 where

$$b_2 \equiv a_2 - d + d' \pmod{c}$$

It is easy to see that the ith person, having heard the $i -$ 1 people behind him and seeing the hats in front of him, has enough information to deduce his hat color in the same way. Hence only the first person might get it wrong.

End of Solution Two

17.6 One Solution to the Infinite Colors, Infinite People Hat Problem

The solution to the infinite colors, infinite people hat problem is not just similar to the c colors, infinite people, it's actually identical! We leave it to the reader to check this.

17.7 A Point About the Infinite Colors, Infinite People Hat Problem

The solution to the infinite people hat problem partitions students into four groups:

- Those that understand it and like it. These are usually students who are interested in pure math and hence do not care that the solution cannot really be carried out.
- Those that understand it but don't like it. These are very good students who like practical things. I've gotten the complaint *but nobody could memorize all of those representatives*. They are, of course, correct. But these problems aren't really practical anyway!

- Those that do not understand it but like it. They see that something interesting is going on and may later understand it.
- You can guess.

17.8 When I DUMPED MY TABLES

The story I told in Section 17.2 left out a word. I didn't tell my students *DUMP YOUR TABLES*. I told them *DUMP YOUR F**KING TABLES*. I said it a lot.

A TRUE DIALOGUE:

Bill: I am going to write a blog about the hat problem and those students and title it *DUMP YOUR F**KING TABLES*.

Clyde: If use the expletive it will detract from the point of the post and therefore you should drop it.

Bill: The expletive is good for emphasis and is part of the story.

Clyde: It is more distracting than helpful.

Bill: But I like it.

Clyde: It is time for you to DUMP YOUR F**KING TABLES!

Bill: Good point.

Chapter 18

Trick Question or Stupid Question?

Prior Knowledge Needed: None, but be willing to be tricked!

18.1   Point

When someone gives you a question or puzzle there may be a trick to solving it. Sometimes you figure out the trick. Other times you are told what the trick is and say *Oh, that's neat!* Other times you say *Really? You've got to be kidding! That's just stupid!* In those cases you declare that it was a stupid question, not a trick question.

I present several questions. You can work on them and decide if you think they are trick questions or stupid questions. There are no right answers to that question, but it's fun to think about. Try to avoid the usual fallacy:

- If I see the trick then it's a trick question.
- If I don't see the trick then it's a stupid question.

I later give the common answer (the one you might get if you don't see the trick), the real answer, and some commentary, including whether it is a trick question or a stupid question.

18.2   The Questions

(1) What is the least common birthday in America?

(2) What is the degree of the polynomial

$$(x - a)(x - b)(x - c) \cdots (x - z)?$$

(3) What US state has the easternmost point in America?

(4) What are the least common first names for a US President as of 2018?

(5) What are the next two numbers in the sequence below?

$$2, 4, 6, 30, 32, 34, 36, 40, 42, 44, 46, 50, 52, 54, 56, 60, 62, 64, \ldots$$

(6) An expert on tracking animals notices one day that there are bear tracks and rabbit tracks converging. He can estimate that they met at 6:00PM with a margin of error of 17 seconds. Hence they must have been there at the same time. He also notices that from the spot they met only rabbit tracks can seen leaving that point. There are some bear bones in the area. *How can a rabbit eat a bear for dinner?*

(7) You have 6 blue socks, 8 white socks, and 10 black socks in a drawer. You take them out of the drawer randomly, one at a time. How many do you need to take out in order to ensure that you have a pair?

(8) What is the square root of nine? Give the answer as an anagram of the original (seven word) question.

(9) The following quote is from the back of a book that I dusted off and took off my shelf recently: *Fortran is one of the oldest high-level languages and remains the premier language for writing code for science and engineering applications.* What year was the book written? If you come within five years then you get it right.

(10) A bird's stomach can hold ten worms. How many worms can a bird eat on an empty stomach?

(11) True or false: If the powerset of the set A has five elements then A is infinite.

(12) The universe of discourse is the natural numbers, which we will assume does not include 0. How do you compliment the set $\{1, 3, 5, \ldots\}$?

(13) What is the next letter in the following sequence?
$$X,J,U,O,M,J,U,G$$

(14) How many legs would a dog have if we called the dog's tail a leg?

(15) The following are two real conversations. For each one: (1) Is the examiner correct? (2) Where and when do you think this conversation took place?

Conversation 1:

Examiner: What is the definition of a circle?

Student: The set of points equidistant from a given point.

Examiner: Wrong! It is the set of *all* points equidistant from a given point.

Conversation 2:

Examiner: What is the definition of a circle?

Student: It is the set of all points equidistant from a given point.

Examiner: Wrong! You did not specify that the distance is nonzero.

18.3   Answers

(1) What is the least common birthday in America?

COMMON ANSWER: Gee, how should I know?

CORRECT ANSWER: Feb 29.

MISC: If we disallow Feb 29 then its December 25. People tend to not want their kid born on Christmas so they induce earlier or try to delay. The most common birthday is September 16. (New York Times, December 19, 2006.) Nobody seems to know why, but there are a number of theories (Medical Daily):

> *including deterioration of sperm quality during summer, seasonal differences in anterior pituitary-ovarian function caused by changes in the daylight length, and variation in quality of the ovum or endometrial receptivity. Or, it could be the increased sexual activity that comes with end-of-the-year festivities.*

COMMENTARY: I call this a trick question, not a stupid question.

(2) What is the degree of the polynomial

$$(x - a)(x - b)(x - c) \cdots (x - z)?$$

COMMON ANSWER: 26.

CORRECT ANSWER: 0. Note that the third to last factor is 0.

COMMENTARY: I call this a trick question, not a stupid question. Most people who see this for the first time (including me) are tricked, but then enlightened and amused when they see the real answer.

Several anonymous commenters pointed out that some people define the degree of the 0 polynomial to be $-\infty$ so that the following equation always holds: if p and q are polynomials then $\deg(pq) = \deg(p) + \deg(q)$.

An anonymous commenter liked this question: *Very interesting trick question, brings up all kinds of issues of type safety and lack of specificity in mathematics educa-tion.* However, a type theorist I showed this to did not like it. He got very angry and told me the degree is 26 because of issues of types-of-variables. That is the 'x' in $(x - a)$ is of a different type then the second x in $(x - x)$. He would call it a stupid question. I would call him a sore loser.

(3) What US state has the easternmost point in America?

COMMON ANSWER: Maine (Quoddy Head Maine).

CORRECT ANSWER: Alaska (Cape Wrangell Alaska). The Prime Meridian is just east of Alaska's Islands so technically (and only technically) the easternmost point in America is in Alaska.

COMMENTARY: This question was one of two questions that inspired this chapter. (The socks question is the other.) My opinion — this is stupid. Just because the Prime Meridian happens to barely split Alaska's Islands does not make Alaska partly in the east. The answer really should be Maine. Peter Winkler [Winkler (2003)] agrees, and his book was my original source.

CLYDE'S COMMENTARY: I do NOT think this is stupid. Why should the technically correct answer to a question not be the real answer? In fact it brings up an interesting point of how to ask the question we were expecting in a natural way.

(4) What are the least common first names for a US President as of 2018?

COMMON ANSWER ONE: Barack. Most people assume, correctly, that there has only been one president with first name Barack. Others are still amazed that America elected a president named Barack.

COMMON ANSWER TWO: Take all of the presidents whose first name no other president has. All of these first names are tied for least common first name. Here are all such presidents in chronological order: Thomas Jefferson, Martin Van Buren, Zachery Taylor, Millard Fillmore, Abraham Lincoln, Ulysses Grant, Rutherford Hayes, Chester Author, Grover Cleveland, Benjamin Harrison, Theodore Roosevelt, Woodrow Wilson, Warren Harding, Calvin Coolidge, Herbert Hoover, Harry Truman, Dwight Eisenhower, Lyndon Johnson, Richard Nixon, Gerald Ford, Ronald Reagan, Barack Obama, and Donald Trump.

Note that Tom, Dick, and Harry, the generic common first names, are represented only once each. I would have thought that one of those names would be more common as an American President's first name than Barack.

CORRECT ANSWER: All of the names that no president had are tied.

COMMENTARY: I think this is a stupid question. When the question is asked there is an implicit assumption that the universe of discourse is the set of all presidents. So the answer should be Common Answer Two. I got it wrong when first asked.

CLYDE'S COMMENTARY: I think this is a fine (although not particularly clever) question. It is clear that there are going to be too many singletons, so there must be a trick.

MISC TRIVIA: The most common first name is James with James Madison, James Monroe, James Polk, James Buchanan, James Garfield, and James Carter. James Carter is the only president ever to be sworn in by his nickname Jimmy. Had Jeb Bush become president he would have been the second. His real first name is John.

(5) What are the next two numbers in the sequence below?

$2, 4, 6, 30, 32, 34, 36, 40, 42, 44, 46, 50, 52, 54, 56, 60, 62, 64, \ldots$

COMMON ANSWER: 66,70? Not with any confidence.

CORRECT ANSWER: The sequence is all numbers that, when written in English, do not have an e in them. $x = 66$, $y = 2000$.

COMMENTARY: When you are given a sequence of numbers you tend to think that the answer will depend on math. Even so, when most people see the answer they think *Oh, that's neat!* So I'll go with trick question.

CLYDE'S COMMENTARY: Once you have seen a couple of similar questions you can recognize that this is not likely to be a *math* question. At that point the universe of possible solutions is too open-ended for my taste. I'll go with trick but boring; I did enjoy seeing the answer.

(6) An expert on tracking animals notices one day that there are bear tracks and rabbit tracks converging. He can estimate that they met at 6:00PM with a margin of error of 17 seconds. Hence they must have been there at the same time. He also notices that from the spot they met only rabbit tracks can seen leaving that point. There are some bear bones in the area. *How can a rabbit eat a bear for dinner?*

COMMON ANSWER: Gee... I don't know.

CORRECT ANSWER: By skipping breakfast and lunch.

COMMENTARY: This is stupid, but I still laughed at it when I first saw it. Of course, I was young at the time. I was 45.

ANOTHER CORRECT ANSWER: By Rabbit I meant Volkswagen Rabbit. It ran over the bear, killed it, and

took it home for dinner. The driver had skipped breakfast and lunch. This was given to me by Evan Golub after
hearing my stupid answer.

(7) You have 6 blue socks, 8 white socks, and 10 black socks
    in a drawer. You take them out of the drawer randomly,
    one at a time. How many do you need to take out in order
    to ensure that you have a pair?

    COMMON ANSWER: 4.

    CORRECT ANSWER: 2. As soon as you take two out you
    have a pair. The problem does not say that the socks have
    to be the same color.

    COMMENTARY: This question was one of two questions
    that inspired the post. (The one about easternmost point
    in the US was the other.) When I first read it in Christian
    Constanda's book [Constanda (2010)] I was annoyed and
    thought it was stupid. On reflection however I think it
    teaches one to read carefully.

    CLYDE'S COMMENTARY: It is implicit that a pair of
    socks means a matched pair. For example, if you were to
    ask someone to get a you a pair of socks from the drawer,
    you would expect to receive two socks of the same color.
    It is a close call, but I think it is stupid.

(8) What is the square root of nine? Give the answer as an
    anagram of the original (seven word) question.

    COMMON ANSWER ONE: 3, but I can't make it into an
    anagram

    COMMON ANSWER TWO:

    > The numerical answer is three
    > But goodness golly gee
    > I can't make it an an-a-gram

I find myself in-a-jam
But I know the answer is three

CORRECT ANSWER: Three, for an equation shows it.

ANOTHER CORRECT ANSWER: The square root of nine is what. This was given to me by Natalie Collina after she heard the answer and thought she could do better. She did!

COMMENTARY: I don't think this is a stupid question, but it is an odd question.

CLYDE'S COMMENTARY: If you asked the question then the answer might be an anagram of "the original (seven word) question".

(9) The following quote is from the back of a book that I dusted off and took off of my shelf recently: *Fortran is one of the oldest high-level languages and remains the premier language for writing code for science and engineering applications.* What year was the book written? If you come within 5 years then you get it right.

COMMON ANSWER: Fortran has not been *the premier language for writing code for science and engineering applications* for a very long time, and you had to dust it off, so I would say 1972. I would not say the 1960's since the quote said that Fortran has been around for a long time.

CORRECT ANSWER: The book is *Modern Fortran* by Clerman and Spector. It came out in 2012. I need to dust my shelves more often.

COMMENTARY: Some people guessed correctly that the book was recent since *if it were an old book you wouldn't be asking this.* Putting that aside, what is the state of Fortran now? Comments on my blog post, plus asking around, indicate that Fortran is still used some and there is of course much legacy code, but it is clearly not *the premier language for writing code for science and engineering applications.* So to get the question right one would have to realize that what was written on the back of the book was not correct. Hence this is a stupid question.

(10) A bird's stomach can hold 10 worms. How many worms can a bird eat on an empty stomach.

COMMON ANSWER: 10 worms.

REAL ANSWER: Once the bird eats one worm the birds stomach is no longer empty.

OBJECTION: Once the bird eats ϵ of the worm his stomach is no longer empty. So the question may be ill defined.

COMMENTARY: I think this question relies on verbal trickery so I would say the question is stupid. Plus the caveat about ϵ makes it a bad question.

(11) True or false: If the powerset of the set A has 5 elements then A is infinite.

COMMON ANSWER: False since the powerset of A cannot be five.

REAL ANSWER: True since the powerset of A cannot be five. The statement is true vacuously. If you want to show that it is false you need to find a set A whose powerset is of size 5 and is finite. Since the size of the powerset of a finite set is always a power of two, there are no sets with powerset 5.

COMMENTARY: This is a trick question but not a stupid question. I did an experiment with this question. I put the question on my Discrete Math final as one of many *True-False-No-Explanation-Needed* questions, and, a couple of years later, put the same question on my Discrete Math final as one of many *True-False-and-Explain-your-answer* questions. My thought was that if a student didn't have to explain the answer they would fall for the trick and answer FALSE, but if they had to explain it, they would rethink it and realize that the statement was TRUE. This was not the case. On the "don't-explain" exam 14 out of 150 got it right. Of the "explain" exam only 19 out of 152 got it right. Even more telling, there was very little correlation between who got it right and how they did on the rest of the exam, on their final grade, or if they were in the honors section. Hence not only was my null hypothesis wrong, it was an awful exam question. I suspect my students thought it was a stupid question.

(12) The universe of discourse is the natural numbers, which we will assume does not include 0. How do you compliment the set $\{1, 3, 5, \ldots, \}$?

COMMON ANSWER: $\{2, 4, 6, \ldots, \}$

REAL ANSWER: There are many ways to do this. One is *My set, you look lovely today!* Or *I think the fact that all your elements are odd makes you nicely proportioned.* Note that I said *compliment*, which is to say something nice, not *complement*.

COMMENTARY: If this question is in written form then someone might catch the spelling difference and get it right. If this question is stated out loud then it is really hard to get it right since (1) *complement* and *compliment* sounds so similar, and (2) the student is not expecting it. I pull this joke on my students about twice a semester. The first time they fall for it; however, the second time, they still fall for it. *Fool me once, shame on you, fool me twice shame on me.*

CLYDE'S COMMENTARY: Bill's blog posts are known for their many spelling errors (along with their random capitalizations). It is ironic that he has a puzzle based on spelling a word correctly.

BILL'S COMMENTARY ON CLYDE'S COMMENTARY: Indeed! People reading the blog did misread it assuming it was just another bad spelling. If one gets it wrong for that reason that's not quite fair.

(13) What is the next letter in the following sequence?
X,J,U,O,M,J,U,G

COMMON ANSWER: Gee, I don't know!

REAL ANSWER: Look at the sentence. *What is the next letter in the following sequence?* If you take the first letter of each of the words you get W,I,T,N,L,I,T,F,S. If you advance each of those by one letter you get X,J,U,O,M,J,U,G,T. Hence the answer is T.

COMMENTARY: This is not a trick question but its a hard question.

CLYDE'S COMMENTARY: This is clearly not a *math* question, so it is fine. It is neither a trick question nor a stupid question. It is a boring question.

(14) How many legs would a dog have if we called the dog's tail a leg?

There are two different answers to this question. I respect both of them.

(a) The answer is clearly five — since 4+1=5. Duh.
(b) Calling the tail a leg does not make it a leg. A dog has four legs. Duh. Consider the following version: If I called all six presidents with first name James, *Alaskans*, then how many are Alaskan? Still zero.

I have seen people on either side not be able to even understand the other side's point. This question has been attributed to Abraham Lincoln; however, just as Bogart never said *Play it again Sam* and Kirk never said *Beam me up Scotty*, Lincoln may never have said calling a tail a leg does not make it a leg (though he might have said *Play it again Sam* or *Beam me up Scotty*).

Here is how the story goes: Lincoln was pondering the Emancipation proclamation and wondered if the president has the authority to issue it. He was concerned that just because he calls a slave a free man does not make him a free man. He used the dog-question to make that point.

In our terminology, Lincoln's point was that definitions should conform to reality and if they do not then it is the definitions that are wrong. I suspect he would not have liked the Banach-Tarski Paradox.

The puzzle was actually around well before Lincoln. I've seen it refer to sheep rather than dogs.

How would you answer the question? Do you understand the opposing viewpoint?

(15) The following are two real conversations. For each one: (1) Is the examiner correct? (2) Where and when do you think this conversation took place?

Conversation 1:

Examiner: What is the definition of a circle?

Student: The set of points equidistant from a given point.

Examiner: Wrong! It is the set of *all* points equidistant from to a given point.

Conversation 2:

Examiner: What is the definition of a circle?

Student: It is the set of all points equidistant from a given point.

Examiner: Wrong! You did not specify that the distance is nonzero.

COMMON ANSWER FROM MATHEMATICIANS: The examiner is wrong. He is being pedantic and formal on an absurd level.

AN ANSWER I HAVE HEARD FROM SOME NON-MATHEMATICIANS: Since mathematicians are often pedantic and formal to an absurd level the examiner is correct.

REAL ANSWER: Nobody in math would be that pedantic. So where and when did this happen? In the old USSR, entrance exams for Moscow State University were rigged so that Jewish students could not pass. The following is a quote from George Szpiro [Szpiro (2007)]. The first story in it happened to Edward Frenkel when he was a 16-year-old taking the oral entrance exam to Moscow State University in 1984 [Frenkel (2013)].

> *Jews — or applicants with Jewish-sounding names — were singled out for special treatment. On one occasion a candidate was failed for answering the question, what is the definition of a circle with the set of points equidistant to a given point. The correct answer, the examiner said, was the set of all points equidistant to a given point. On another occasion an answer to the same question was deemed*

incorrect because the candidate had failed to stipulate that the distance had to be nonzero.

A different technique used on the entrance exams was to give Jewish students problems that had simple solutions which were extremely difficult to find. The simplicity of the solution made appeals and complaints difficult. Some of these problems and their history is in an article by Tanya Khovanova and Alexey Radul [Khovanova and Radul (2012)].

References

Constanda, C. (2010). *Dude, can you count? Stories, challenges, and adventures in mathematics* (Copernicus).

Frenkel, E. (2013). *Love & Math: The heart of hidden reality* (Basic Books).

Khovanova, T. and Radul, A. (2012). Killer problems, *The American Mathematical Monthly* **119**, pp. 815–829, url-https://arxiv.org/abs/1110.1556.

Szpiro, G. (2007). Bella Abramovna Subbbotovskaya and the "Jewish People's university", *NAMS* **54**, pp. 1326–1330, http://www.ams.org/notices/200710/tx071001326p.pdf.

Winkler, P. (2003). *Mathematical Puzzles: A Connoisseur's Collection* (A.K. Peters).

Chapter 19

Multiparty Communication Complexity: A Fun Approach

Prior Knowledge Needed: None. Which is why it's fun!

19.1 Points

Sometime we work on problems that a civilian *could understand and find FUN*, but the solution would be *neither understandable nor FUN*. Perhaps a weaker solution could be found that would be fun!

We present a problem in Multiparty Communication Complexity and then a FUN, but far-from-optimal solution. In the last section we give the history of the problem and the known results. The optimal results require Ramsey Theory, and hence are not likely to be found by many civilians.

19.2 A FUN Problem in Multiparty Communication

Alice, Bob, and Carol each have an integer between 0 and $2^n - 1$ (inclusive) on their forehead, written in binary; so everyone has exactly n bits on their forehead. Everyone can see the two numbers that are NOT on their own forehead. Call the three numbers a, b, c. They wish to determine if $a + b + c = 2^{n+1} - 1$ (the bit string 1^{n+1}). Each player can shout information that the other two hear. At the end of the protocol they should all know the answer.

There is an easy solution: Alice shouts y. Then Bob (or Carol) can compute $a+b+c$ and shout YES if $a+b+c = 2^{n+1}-1$ and NO if not. This protocol takes $n+1$ bits of communication. Can they do this with fewer than $n + 1$ bits?

The problem *sounds* like it could be FUN. The best protocol known involves 3-free sets, which arise from lower bounds on van der Warden numbers. That is FUN for fans of Ramsey Theory, but not FUN for most. Is there a FUN solution to this problem where the solution uses fewer than n bits of communication?

What is FUN? We can't define it rigorously but some characteristics are (1) uses only elementary mathematics, (2) a civilian has a good chance of solving it, (3) if a civilian does not get it right and we show her the solution then she will kick herself for not solving it.

The answer is on the next page! Try to figure it out yourself before going there.

19.3 A FUN Solution

I posted the following question on my blog:

is there a FUN solution with communication $< n$ bits

I got the following answer from commenter Dean Foster:

Theorem 19.1. *There is a protocol for this problem where $\frac{n}{2} + O(1)$ bits are communicated.*

Proof. We assume n is even. We leave it to the reader to modify the proof for the case where n is odd.

Alice has $a = a_{n-1} \cdots a_0$. Bob has $b = b_{n-1} \cdots b_0$. Carol has $c = c_{n-1} \cdots c_0$.

Alice shouts the following sequence of bits:

$$b_{n-1} \oplus c_0, b_{n-2} \oplus c_1, \cdots, b_{n/2} \oplus c_{n/2-1}.$$

Note that this is $n/2$ bits. Since Bob knows all of the c_i he now knows $b_{n/2}, \ldots, b_{n-1}$. Since Carol knows all of the b_i she now knows $c_0, \ldots, c_{n/2-1}$.

Carol knows $a_0, \ldots, a_{n/2-1}$, $b_0, \ldots, b_{n/2-1}$, $c_0, \ldots, c_{n/2-1}$. Hence she can compute the sum of the lower $n/2$ bits of a, b, c:

$$a_{n/2-1}a_{n/2-1} \cdots a_0 + b_{n/2-1}b_{n/2-1} \cdots b_0 + c_{n/2-1}c_{n/2-1} \cdots c_0.$$

The players think of this as an $(n/2)$-bit string and a carry bit. If the string is $1^{n/2}$ then Carol shouts YES along with the carry bit. If the string is not $1^{n/2}$ then Carol shouts NO and the protocol ends with everyone knowing $a + b + c \neq 2^{n+1} - 1$. Note that Carol shouts at most 2 bits.

Assume Carol shouts YES and the carry bit. Bob knows $a_{n/2}, \ldots, a_{n-1}$, $b_{n/2}, \ldots, b_{n-1}$, $c_{n/2}, \ldots, c_{n-1}$ and the carry bit. Hence Bob can compute $a + b + c$. If the answer is $2^{n+1} - 1$ then he shouts YES, else he shouts NO. Note that Bob shouts 1 bit.

The total number of bits shouted is $\frac{n}{2} + O(1)$. $\quad\square$

19.4 A Sequel You Can Ask a Civilian

If you present this problem to a civilian then one of the following might happen:

(1) She cannot solve it and does not think it's FUN. Oh well.
(2) She solves it and does not think it's FUN. Really? This would be rare.
(3) She cannot solve it but thinks it's FUN.
(4) She solves it and thinks it's FUN.

We are interested in the cases where she thinks it's FUN. In this case (1) if she didn't solve it then show her the solution, and (2) give her the following problems:

(1) Alice, Bob, Carol, and Donna each have a number between 0 and $2^n - 1$ on their forehead, written in binary; so everyone has exactly n bits on their forehead. Everyone can see the three numbers that are NOT on their own forehead. Call the three numbers a, b, c, d. They wish to determine if $a + b + c + d = 2^{n+1} - 1$. There is an obvious $(n+1)$-bit protocol. Can they do better? Try to do better than $\frac{n}{2} + O(1)$.

(2) Generalize this to k people, $A_1, \ldots, A_k$. They each have an integer number between 0 and $2^n - 1$ (inclusive) on their forehead, written in binary; so everyone has exactly n bits on their forehead. Everyone can see the $k - 1$ numbers that are NOT on their own forehead. Call the k numbers $a_1, \ldots, a_k$. They wish to determine if $a_1 + \cdots + a_k = 2^{n+1} - 1$. There is an obvious $(n+1)$-bit protocol. Can they do better? Try to make the complexity go down as k goes up.

19.5 Is the Problem Fun? An Experiment

In the Summer of 2018, I ran an REU program with 32 undergrads. The undergrads were working on projects involving theory so they were good at math. We sometimes had them work on FUN (we hope) math problems over lunch. For one of those lunches we gave them the forehead problem to solve (the exact handout is in the Section 19.8).

The students were in five groups of size three to six. Three groups got it right, a fourth overheard the solution, and one group did not get it. Our impression is that $2/3$ of them thought it was FUN.

We asked them to vote on options for a better protocol. Some thought that $\frac{n}{3} + O(1)$ was the best one could do, some thought maybe αn for some $\alpha < \frac{1}{2}$ and a very few thought that $O(\frac{n}{\log n})$ was possible — though only because *you would not make that an option unless it was true.* After the vote one student said

You can do it with $O(\sqrt{n})$ bits by using Ramsey Theory!

He is correct. He had my course in Ramsey Theory the prior semester and just now remembered that he had seen the problem before. Since I presented it as a FUN problem to do over lunch he forgot that it was a problem that he had learned about in a more serious way.

19.6   Multiparty Communication Complexity: A Serious Approach

In 1983 Chandra *et al.* [Chandra *et al.* (1983)] introduced multiparty communication complexity in order to get better bounds on constant width branching programs. They wanted to get a non-polynomial lower bound on the length; however, they only got a very weak *non-linear* lower bound. A later paper by Ajtai *et al.* [Ajtai *et al.* (1986)] improved this to an $\Omega(n \log n)$ lower bound. However, Barrington [Barrington (1989)] showed that their goal was impossible.

Let f_3 be the problem in Section 19.2 and f_k be the problem in Section 19.4. Let $d(f_3)$ be the number of bits communicated in the best protocol for f_3. Similar for f_k.

Chandra *et al.* proved:

(1) $d(f_3) \leq \sqrt{\log n}$
(2) for all $k \geq 3$, $d(f_k) \geq \Omega(1)$.

Beigel *et al.* [Beigel *et al.* (2006)] proved:

(1) $d(f_3) \geq \Omega(\log \log n)$.
(2) $d(f_k) \leq O(n^{1/(\log_2(k-1)+1)})$.

The proofs of the upper and lower bounds on both the 3-person and k-person case used results from Ramsey Theory. I asked Ramsey theorists about what is believed and how it would relate to this problem. The consensus is that (1) the upper bounds are likely the answer, and (2) there will be no improvement on the lower bounds for quite some time.

19.7   Open Problems

Another student raised the following question: Is there a slightly harder protocol that gives a slightly better protocol, say $\frac{n}{3}+O(1)$ bits. Or even $O(\frac{n}{\log n})$. So an intermediary between the $\frac{n}{2}+O(1)$ and $O(\sqrt{n})$ solutions in both quality and hardness. As far as I know, the answer is no. In other words:

Open: Is there an elementary protocol using at most $\alpha n + O(1)$ bits where $\alpha < \frac{1}{2}$? Is there an elementary protocol that is sublinear?

The upper bound is a fine problem. A lower bound would be very hard to even state since *elementary protocol* is not well defined.

19.8   The Original Worksheet

Alice, Bob, and Carol each have a number between 0 and $2^n - 1$ on their forehead, written in binary; so they each have an n-bit string on their forehead. Everyone sees the two numbers that are NOT on their own forehead. The numbers are a, b, c. They want to know if $a+b+c = 2^{n+1}-1$. Each player can shout information. At the end of the protocol they should all know the answer.

Here is a protocol: Alice shouts a. Then Bob computes $a+b+c$. He shouts YES if $a+b+c = 2^{n+1} - 1$ and NO if not. This protocol takes $n + 1$ bits of communication. Can they do this with fewer than $n + 1$ bits?
YES!
There is a way to do this with $\frac{n}{2} + O(1)$

We chose to tell them the answer since otherwise it might be too hard. We were right — it was hard enough as it was.
Try to find it!

Generalize to four People! k People!
Alice, Bob, Carol, and Donna each have a number between 0 and $2^n - 1$ on their forehead, written in binary. Call the four numbers a, b, c, d. They wish to determine if $a + b + c + d = 2^{n+1} - 1$. Try to get a protocol that works in $\frac{n}{3} + O(1)$ bits.

$A_1, \ldots, A_k$ each have a number between 0 and $2^n - 1$ on their forehead, written in binary; so everyone has exactly n bits on their forehead. Call the k numbers $a_1, \ldots, a_k$. They wish to determine if $\sum_{i=1}^{k} a_i = 2^{n+1}-1$. Try to get a protocol that works in $\frac{n}{k-1} + O(1)$ bits.

Think About

For the three-player version, we got a protocol with $\frac{n}{2} + O(1)$ bits.

Is there a better protocol?

Is there a protocol with $\frac{n}{3} + O(1)$ bits?

Is there a protocol with $O(\frac{n}{\log n})$ bits?

Is there a protocol with $O(n^{0.99})$ bits?

Can you show that any protocol requires BLAH bits?

References

Ajtai, M., Babai, L., Hajnal, P., Kolmos, J., Pudlak, P., Rodl, V., Szemeredi, E., and Turán, G. (1986). Two lower bounds on branching programs, in *Proceedings of the Eighteenth Annual ACM Symposium on the Theory of Computing*, Berkeley CA, pp. 30–38.

Barrington, D. (1989). Bounded-width polynomial-size branching programs recognize exactly those languages in NC^1, *Journal of Computer and System Sciences* **38**, pp. 150–164.

Beigel, R., Gasarch, W., and Glenn, J. (2006). The multiparty communication complexity of exact-t: improved bounds and new problems, in *Proceedings of the 31th International Symposium on Mathematical Foundations of Computer Science 2001*, Stara Lesna, Slovakia, pp. 146–156.

Chandra, A., Furst, M., and Lipton, R. (1983). Multiparty protocols, in *Proceedings of the Fifteenth Annual ACM Symposium on the Theory of Computing*, Boston MA, pp. 94–99, http://portal.acm.org/citation.cfm?id=808737.

PART 3

Theorems with a Point

Chapter 20

What is Obvious? Less Elegant Proofs That $\binom{2n}{n}$ and $C_n = \frac{1}{n+1}\binom{2n}{n}$ are Integers

Prior Knowledge Needed: None.

20.1 Point

A conversation between Bill and Olivia (the 14-year old daughter of a friend).

Bill: Do you know what 10! is?

Olivia: Yes, when I turned 10 I yelled out **I AM 10!**

Bill: I mean it in a different way. In math 10! is $10 \times 9 \times 8 \times 7 \times 6 \times 5 \times 4 \times 3 \times 2 \times 1$.

Olivia: Okay. So what? And why that "×1?" Thats just stupid.

Bill: The "×1" is traditional. Do you think that $\frac{10!}{5!5!}$ is an integer?

Olivia: No but I'll work it out and see.
$$\frac{10!}{5!5!} = \frac{10 \times 9 \times 8 \times 7 \times 6 \times 5!}{5!5!} = \frac{10 \times 9 \times 8 \times 7 \times 6}{5 \times 4 \times 3 \times 2}$$

$$= \frac{10}{5 \times 2} \times \frac{9}{3} \times \frac{8}{4} \times 7 \times 6 = 3 \times 4 \times 7 \times 6.$$

Pierce my ears and call me drafty! It IS an integer!

Bill: Do you think that $\frac{100!}{50!50!}$ is an integer?

Olivia: Fool me once shame on you, Fool me twice... uh, uh, who sang *Won't get fooled again?*

Bill: Yes, and they were great!

Olivia: But... who sang it?

Bill: Naturally, cover bands often sing it also.

Olivia: Which band are they covering?

Bill: [The] Who — but never mind all of that, we can insert a *Who's on First* imitation in the next edition. Back to the problem.

Olivia: What?

Bill: He's on second, but never mind that. It turns out that (1) $\frac{100!}{50!50!}$ is an integer, and (2) I can prove it without actually calculating it.

Olivia: How?

Bill: How many ways can you order n elements?

Olivia: There are n choices for the first spot, $n - 1$ for the second spot, etc. OH, it's that $n!$ thing. Cool!

Bill: What if you only wanted k of the items?

Olivia: There are n choices for the first spot, $n - 1$ for the second spot, $n - 2$ for the third, etc., but you stop at $(n - k + 1)$ for the kth spot. So thats $n \times n - 1 \times \cdots \times n - k + 1$. OH, thats $\frac{n!}{(n-k)!}$. Cool!

Bill: What if you want to choose k elements but don't care about the ordering?

Olivia: Well, if you DO care about the ordering, then it's
$\frac{n!}{(n-k)!}$. So if you list out all of those but don't want
to count different permutations of the k elements
as different, then you would divide by $k!$. So it's
$\frac{n!}{(n-k)!k!}$.

Bill: Yes! And note that the answer must be an integer.
If $n = 100$ and $k = 50$ then $\frac{100!}{50!50!}$ is the answer.
Hence it is an integer.

Olivia: You call that a proof? That's INSANE! You can't
just solve a problem that must have an integer solution and turn that into a proof that the answer
is an integer! Its unnatural! It is counter to the
laws of God and Man! AH — I have a counterex-
ample! By the same reasoning $\frac{97!}{43!54!}$ is an integer
when clearly it can't be. So there!

Bill: Work it out and you'll see that it's an integer.

Olivia: (She does). Oh. Hmmm. Well, maybe the proof is
correct after all. But I still don't like it.

Inspired by Olivia, I came up with a LESS elegant proof that
$\frac{(2n)!}{n!n!}$ is always an integer. Inspired by that, I also came up with
a LESS elegant proof that $\frac{1}{n+1}\frac{(2n)!}{n!n!}$, the Catalan numbers, are
integers.

If you are reading this book then you probably think it is
OBVIOUS that $\frac{n!}{k!(n-k)!}$ is an integer. Those of you who know
what the Catalan numbers are think that it is OBVIOUS they
are integers. You are right that these are integers; however,
there was a time when you might have reacted like Olivia, or at
least thought it bizarre that this is a proof of integrality.

When presenting mathematics we should be aware that what
we believe is obvious comes from years of exposure that our
audience may not have.

20.2   Notation

Notation 20.1. $\binom{n}{k}$ is $\frac{n!}{k!(n-k)!}$. C_n is $\frac{1}{n+1}\binom{2n}{n}$.

We give several proofs that $\binom{2n}{n}$ is an integer, none of them combinatorial. We (1) prove that C_n solves a combinatorial problem, hence is an integer, and (2) give one non-combinatorial proof that C_n is an integer.

20.3   Proof by Algebra and Induction That $\binom{2n}{n}$ is an Integer

By an algebraic argument one can show that, for $1 \le k \le n$,

$$\binom{n}{k} = \binom{n-1}{k-1} + \binom{n-1}{k}.$$

Using this one can show, by induction on n, that $\binom{n}{k}$ is an integer. Hence $\binom{2n}{n}$ is an integer.

This proof demonstrates the principle that it is sometimes easier to prove a harder theorem. We just wanted $\binom{2n}{n}$ to be an integer and we ended up with the more general theorem that $\binom{n}{k}$ is an integer. Is there a proof by induction that $\binom{2n}{n}$ is an integer that does not use $\binom{n}{k}$? I do not know of one.

20.4   Proof by Number Theory That $\binom{2n}{n}$ is an Integer

Let p be a prime. We want to find x and y such that

(1) x is the largest number such that p^x divides $(2n)!$
(2) y is the largest number such that p^y divides $n!n!$

We then show that $x \ge y$ to complete the proof.

The number of times the factor of p occurs in $m!$ is

$$\left\lfloor \frac{m}{p} \right\rfloor + \left\lfloor \frac{m}{p^2} \right\rfloor + \left\lfloor \frac{m}{p^3} \right\rfloor + \cdots$$

Hence

$$x = \left\lfloor \frac{2n}{p} \right\rfloor + \left\lfloor \frac{2n}{p^2} \right\rfloor + \left\lfloor \frac{2n}{p^3} \right\rfloor + \cdots$$

$$y = 2 \left\lfloor \frac{n}{p} \right\rfloor + 2 \left\lfloor \frac{n}{p^2} \right\rfloor + 2 \left\lfloor \frac{n}{p^3} \right\rfloor + \cdots$$

To obtain $x \geq y$ it suffice to show that, for all real α, $\lfloor 2\alpha \rfloor \geq 2 \lfloor \alpha \rfloor$. If α is an integer we get equality. If $n < \alpha < n+1$ then $2n < 2\alpha < 2n+2$ and hence $\lfloor \alpha \rfloor = n$ and $\lfloor 2\alpha \rfloor \geq 2n$. Hence we have $\lfloor 2\alpha \rfloor \geq 2 \lfloor \alpha \rfloor$.

20.5   Combinatorial Proof That C_n, the nth Catalan Number, is an Integer

This is well known, but we include it for completeness.

Let C_n be the number of ways to parenthesize $X \cdots X$ ($n+1$ times). (Note that we use n pairs of parentheses.) The parenthesizing should be such that if the operation is non-associative, the answer is unambiguous.

Example 20.1.

(1) $n = 0$. There is only one way to parenthesize X and that is by X, so $C_0 = 1$.

(2) $n = 1$. There is only one way to parenthesis XX namely XX so $C_1 = 1$.

(3) $n = 2$. XXX can be parenthesized in any of the following ways: $X(XX), (XX)X$. Hence $C_2 = 2$.

(4) $n = 3$. XXXX can be parenthesized in any of the following ways: $((XX)X)X, (X(XX))X, (XX)(XX), X((XX)X), X(X(XX))$. Hence $C_3 = 5$.

We derive a recurrence for C_n. Assume $n \geq 3$. In any parenthesizing of X^n there will Y, Z (both nonempty) such that X^n is parenthesized $(Y)(Z)$ where Y, Z are also parenthesized. Hence, C_n can be defined by $C_0 = 1$ and, for all $n \geq 1$,

$$C_{n+1} = \sum_{i=0}^{n} C_i C_{n-i}$$

Let $C(x) = \sum_{i=0}^{\infty} C_n x^n$. From the recurrence one can show that

$$C(x) = 1 + x C(x)^2.$$

Hence

$$C(x) = \frac{1 - \sqrt{1 - 4x}}{2x} = \frac{2}{1 + \sqrt{1 - 4x}}.$$

From this, the Taylor series yields the result $C_n = \frac{1}{n+1}\binom{2n}{n}$. Since C_n solves a combinatorial problem, C_n is an integer.

If all you wanted to know was that C_n is an integer, would the proof above be elegant? I leave this as a matter of taste.

20.6 Proof by Number Theory That C_n is an Integer

An easy algebraic proof shows $C_n = \binom{2n}{n} - \binom{2n}{n+1}$. By Section 20.4 we already have a proof that $\binom{2n}{n}$ is an integer that just uses number theory. It can easily be modified to show that $\binom{2n}{n+1}$ is an integer. Hence C_n is an integer.

20.7 Open Problem

Is there a proof that C_n is an integer that is similar to the proof in Section 20.4 that $\binom{2n}{n}$ is an integer?

Chapter 21

Is There a Pattern? Is it Interesting? Perfect Numbers and Sums of Cubes

Prior Knowledge Needed: You need to care about perfect numbers.

21.1 Point

Often in math you spot a pattern based on finite evidence and the question arises: *Is it really a pattern?* Once you've proven the pattern persists one should ask *is the pattern interesting?*

Here is a silly example: I notice the following about the primes ≥ 3:

$3 = 2 \times 1 + 1$
$5 = 2 \times 2 + 1$
$7 = 2 \times 3 + 1$

Ah, maybe the ith prime (not including 2) is of the form $2i + 1$. Alas, no, $2 \times 4 = 1 = 9$ which is not prime. But we do note that

$11 = 2 \times 5 + 1$

AH, I conjecture that every prime larger than 3 is of the form $2x + 1$. This is true! This is also not interesting since every odd is of that form and all primes ≥ 3 are odds. In other words, the conjecture (really an obvious fact) is not interesting since its not about primes.

We give an example of finite evidence for a pattern and ask

both questions: *Is there a pattern?* And *is the pattern interest-ing?*

21.2   The Problem

Definition 21.1. $\sigma(n)$ is the sum of the divisors of n including both 1 and n. A number n is *perfect* if $\sigma(n) = 2n$.

It is known that there are an infinite number of even perfect numbers. Theorem 21.1 characterizes the even perfect numbers. It is unknown if there are *any* odd perfect numbers.

The first four perfect numbers are

6

$28 = 1^3 + 3^3$

$496 = 1^3 + 3^3 + 5^3 + 7^3$

$8128 = 1^3 + 3^3 + \cdots + 15^3$

Is there something interesting going on here? Is every perfect number greater than 6 of the form

$$1^3 + 3^3 + \cdots + (2k + 1)^3$$

for some k?

We prove that if n is an even perfect number than there exists k such that n is the sum of the first k odd cubes. We then discuss if this is interesting or not.

21.3   Needed Theorems

Lemma 21.1.

(1) $\sigma(ab) = \sigma(a)\sigma(b)$.
(2) For all x, $\sigma(2^x) = 2^{x+1} - 1$.

Proof. We denote the ith prime by p_i.
1) Let $a, b \in \mathbb{N}$. Let n be such that p_n is the largest prime that divides ab. Let $a = \prod_{i=1}^{n} p_i^{a_i}$ and $b = \prod_{i=1}^{n} p_i^{b_i}$ (some of the a_i

and b_i might be zero).

$$\sigma(a) = \sum_{j_1=0}^{a_{i_1}} \sum_{j_2=0}^{a_{j_2}} \cdots \sum_{j_n=0}^{a_{j_n}} \prod_{j=1}^{n} p_i^{a_{j_i}}$$

$$\sigma(b) = \sum_{j_1=0}^{b_{j_1}} \sum_{j_2=0}^{b_{i_2}} \cdots \sum_{j_n=0}^{b_{j_n}} \prod_{j=1}^{n} p_i^{b_{j_i}}$$

$$\sigma(ab) = \sum_{j_1=0}^{a_{j_1}+b_{j_1}} \sum_{j_2=0}^{a_{i_2}+b_{i_2}} \cdots \sum_{j_n=0}^{a_{j_n}+b_{j_n}} \prod_{j=1}^{n} p_i^{a_{j_i}+b_{j_i}}$$

Algebra establishes the result.

2) Follows from part 1. □

The following is called the Euclid-Euler Theorem since Euclid proved one direction, and Euler the other. In Chapter 5 we proved they never met.

Theorem 21.1. *n is an even perfect number iff there exists p such that $2^p - 1$ is prime and $n = 2^{p-1}(2^p - 1)$. (The number p will also be prime since if $2^p - 1$ is prime then p is prime. We do not need this fact.)*

Proof. 1) (Euler proved this part) If n is an even perfect number then here exists p such that $2^p - 1$ is prime and $n = 2^{p-1}(2^p - 1)$.

Assume n is an even perfect number. Since n is even, there exists $p \geq 2$ and b odd such that $n = 2^{p-1}b$. Since n is perfect

$$\sigma(n) = 2n = 2^p b.$$

By Lemma 21.1

$$\sigma(n) = \sigma(2^{p-1})\sigma(b) = (2^p - 1)\sigma(b).$$

Equating these two different expressions for $\sigma(n)$ we obtain

$$2^p b = (2^p - 1)\sigma(b) \tag{21.1}$$

Since $2^p - 1$ divides $2^p b$ and has no factors in common with 2^p, $2^p - 1$ divides b. Let $b = (2^p - 1)c$. Substituting this expression for b into Equation 21.1 yields

$$2^p(2^p - 1)c = (2^p - 1)\sigma(b)$$

$$2^p c = \sigma(b)$$

We want to show that $c = 1$. Assume, by way of contradiction, that $c \geq 2$. Recall that $\sigma(b)$ is the sum of all the divisors of b including 1 and b. Therefore, since $b = (2^p - 1)c$:

$$\sigma(b) \geq 1 + c + (2^p - 1) + (2^p - 1)c = 2^p(1 + c).$$

Hence

$$2^p c = \sigma(b) \geq 2^p(1 + c)$$

This is a contradiction.

2) (Euclid proved this part) If $n = 2^{p-1}(2^p - 1)$ where $2^p - 1$ is prime then n is perfect.

$$\sigma(n) = \sigma(2^{p-1}(2^p - 1))\sigma(2^{p-1})\sigma(2^p - 1)$$

$$= (2^p - 1)(1 + (2^p - 1)) = 2^p(2^p - 1) = 2 \times 2^{p-1}(2^p - 1) = 2n.$$

$\square$

Note 21.1. For the first perfect number, 6, $p = 2$. Clearly for all later perfect numbers the prime p is odd. This will lead to our main theorem being true for all perfect numbers larger than 6.

Theorem 21.1 reduces the search for even perfect numbers to the search for Mersenne primes, that is, primes of the form $2^p - 1$.

Recall that our concern is sums of consecutive odd cubes. The next theorem characterizes such sums.

Theorem 21.2. *For all* $m \geq 1$, $\sum_{i=0}^{m-1}(2i + 1)^3 = m^2(2m^2 - 1)$.

How you would derive this

We are not going to do a standard proof by induction. Instead we discuss how you might derive this by hand with a minimum of calculation.

Since $\sum_{i=0}^{m-1}(2i+1)^3$ is approximately

$$\int_1^{m-1}(2x+1)^3 dx = 2m^4 + O(m^3)$$

we can guess that the summation is a polynomial of degree 4 with lead term $2m^4$. So we need to find b, c, d, e such that

$$\sum_{i=0}^{m-1}(2i+1)^3 = 2m^4 + bm^3 + cm^2 + dm + e$$

Since when $m = 0$ the sum is 0 we get $e = 0$.

From here there are two ways to proceed: (1) plug in $m = 1, 2, 3$ to get three linear equations in three variables. (2) do a proof by induction and see what the proof forces b, c, d to be. Either technique will derive and prove the theorem.

End of how you would derive this

21.4   The Main Theorem

Based on the examples in Section 21.2, we had pondered if every even perfect number is the sum of an initial segment of odd cubes. The answer is yes! We will now prove this!

Theorem 21.3. *If $n > 6$ is an even perfect number then there exists m such that n is the sum of the first $m - 1$ odd cubes.*

Proof. Let $n > 6$ be an even perfect number. By Theorem 21.1 there exists p such that $2^p - 1$ is prime and $n = 2^{p-1}(2^p - 1)$. By Note 21.1, since $n > 6$, p is odd, so $(p-1)/2 \in \mathbb{N}$.

Let $m - 1 = 2^{(p-1)/2}$. By Theorem 21.2

$$\sum_{i=0}^{m-1}(2i+1)^3 = m^2(2m^2 - 1) = (2^{(p-1)/2})^2(2 \times (2^{(p-1)/2})^2 - 1)$$
$$= 2^{p-1} \times (2^p - 1) = n$$

$\square$

21.5   Is the Theorem Interesting?

Theorem 21.3 *seems* interesting since it gives us a property of perfect numbers. But in the proof we only used that the number n was of the form $2^{p-1}(2^p - 1)$. We *did not* use that $2^p - 1$ is prime. We did use that p is odd (so that $p - 1$ is even). Hence we have the following theorem.

Theorem 21.4. *If n is of the form $2^{p-1}(2^p - 1)$ where p is odd then n is the sum of the first $(p - 1)/2$ odd squares.*

Theorem 21.4 is more general than Theorem 21.3, yet it seems less interesting. Perfect numbers are interesting, but numbers of the form $2^{p-1}(2^p - 1)$ are not. Hence, alas, there was a pattern but it was not interesting. Oh well.

Chapter 22

A Sane Reduction of COL_4 to COL_3

Prior Knowledge Needed: NP-completeness and Reductions.

22.1 Point

Definition 22.1. COL_k is the set of all graphs that are k-colorable.

Karp [Karp (1972)] showed that $\{(G, k) : G \in COL_k\}$ is NP-complete. Is there some fixed k such that COL_k is NP-complete? Yes: Stockmeyer [Stockmeyer (1973)] and Lovász[Lovasz (1973)] independently showed that COL_3 is NP-complete. The gadgets we present later can be used to prove this.

Assume $a < b$. It is easy to show that, $COL_a \leq COL_b$: add K_{b-a} and an edge from every vertex of K_{b-a} to every vertex of G.

Hence we have:
$$COL_3 \leq COL_4 \leq COL_5 \leq \cdots .$$
When I showed $COL_3 \leq COL_4$ in class a student asked:
$$\text{Is } COL_4 \leq COL_3?$$
What do you think? Ponder this before turning to the next page.

22.2 Is $COL_4 \leq COL_3$?

I gave the class three options and asked them to vote:

(1) 10 thought that it is NOT the case that $COL_4 \leq COL_3$.
(2) 12 thought that $COL_4 \leq COL_3$.
(3) 15 thought that the question of $COL_4 \leq COL_3$ is UN-KNOWN TO SCIENCE! (Since the P = NP question is open it is reasonable for the students to think it is UN-KNOWN TO SCIENCE.

I asked one of the NO voters why he thought $COL_4 \leq COL_3$ is not true. He said that he could not think of a reduction, though he admitted this is not a proof. I told him he was WRONG and RIGHT but more WRONG.

By the Cook-Levin Theorem SAT is NP-complete, hence

$$COL_4 \leq SAT.$$

Since COL_3 is NP-complete

$$SAT \leq COL_3.$$

Since reduction is transitive we obtain

$$COL_4 \leq COL_3.$$

So I told the student YES: $COL_4 \leq COL_3$, **but this reduction is insane:** We transform a graph to a formula and the formula back to a graph!

22.3 The Reduction Made Some Students Sad :-(

Later in the semester I asked the students for their favorite and least favorite part of the course. The student who first asked me the question wrote:

My least favorite part was $COL_4 \leq COL_3$. It made me sad.

His ex-girlfriend, who he intentionally sat next to, wrote:

The reduction COL$_4$ $\leq$ COL$_3$ *was the best part of the course since it showed my arrogant ex-boyfriend that just because he can't find a reduction doesn't mean there isn't one.*

In total 8 students said it was their least favorite part of the course, but 10 students said it was their favorite. No other topic did as well in either category.

But I don't like to see my students sad! So I posted the following open problem on my blog:

Is there a sane reduction $COL_4 \leq COL_3$? *More generally, if* $3 \leq a < b$ *is there a sane reduction* $COL_b \leq COL_a$? *Note that we have not, and will not, define* sane *rigorously. We use the same criteria that Supreme Court Justice Potter Stewart used when trying to define obscenity: I know it when I see it.*

My first fear was that someone would give a direct proof (i.e., one that does not use SAT) that used algebraic-geometric-expander-deformation-retractions (I do not know what this means or if such objects exist). This did not happen. However, a reader posted this candidate for a sane reduction:

Let $HCOL_k$ *be the set of all hypergraphs that are k-colorable. Lovász [Lovasz (1973)] showed* $COL_k \leq HCOL_2 \leq COL_3$.

That proof transforms graphs to hypergraphs and back to graphs. Some may regard that as sane. We do not.

My second fear was that nobody would be able to find a reduction. Alas, nobody did. Hence I followed the advice of Charles-Guillaume Étienne:

On n'est jamais servi si bie que par soi-même

which has been translated into

If you want something done right, do it yourself

though it literally means

One is never served so well as by oneself

I found a reduction myself. The next section shows, for all $k \geq 2$, a sane reduction from COL_k to COL_3. I found this reduction too late to make that class happy; however, I have used it ever since to the surprise and delight of my now-happy students.

22.4 Yes There is a Sane Reduction

22.4.1 *The Gadgets Needed*

The following gadget is often used to prove that COL_k is NP-complete.

Definition 22.2. Gadget $GAD(x, y, z)$ is the graph in Figure 22.1. (The vertices that don't have labels are never referred to so we don't need to label them.)

Fig. 22.1　$GAD(x, y, z)$.

We leave the proof of the following easy lemma to the reader.

Lemma 22.1. *If $GAD(x, y, z)$ is three-colored and x, y get the same color, then z also gets that color.*

The number of regions are 1,2,4,8,16, respectively. It is tempting to guess that the number of regions is 2^{n-1}.

Definition 22.3. $GAD(x_1, \ldots, x_k, z)$ is the graph in Figure 22.2 for $k = 5$. Note that we build it out of the gadgets from Figure 22.1. The graph $GAD(x_1, \ldots, x_k, z)$ has $O(k)$ vertices, and $O(k)$ edges.

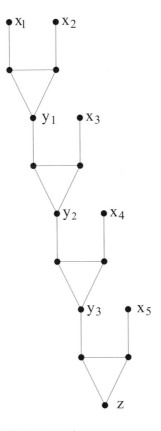

Fig. 22.2 $GAD(x_1, x_2, x_3, x_4, x_5, z)$.

We leave the proof of the following easy lemma to the reader.

Lemma 22.2. *Let* $k \geq 2$. *If* $GAD(x_1, x_2, \ldots, x_k, z)$ *is three-colored and* $x_1, \ldots, x_k$ *get the same color, then* z *also gets that color.*

22.4.2   The Sane Reduction

Theorem 22.1. *Let* $k \geq 2$. $COL_k \leq COL_3$ *by a simple reduction that takes a graph* G *with* n *vertices and* e *edges, and produces a graph* G' *that has* $O(k^2 n + ke)$ *vertices and edges.*

Proof. Let G have vertices $v_1, \ldots, v_n$ and edge set E. We construct G':

(1) There are three special vertices T, F, R, which form a triangle. In any 3-coloring they must have different colors, which we call T, F, RThis is three vertices and three edges.

For $1 \leq i \leq n$ and $1 \leq j \leq k$ there is a vertex v_{ij}. All of these will be connected by an edge to vertex R. Hence each v_{ij} must be colored T or F. This requires be kn vertices and kn edges.

Our intent is: v_{ij} is colored T means that vertex v_i in G is colored j; v_{ij} is colored F means that vertex v_i in G is not colored j. Hence, for all i, there is exactly one j where v_{ij} is colored T.

Here is the formal construction:

(a) For all $1 \leq i \leq n$ we need that *at least one* of $v_{i1}, \ldots, v_{ik}$ is colored T. Hence we need it to not be the case that $v_{i1}, v_{i2}, \ldots, v_{ik}$ are all colored F. We place the gadget $G(v_{i1}, \ldots, v_{ik}, T)$ in the graph. If $v_{i1}, \ldots, v_{ik}$ are all colored F then this gadget will not be 3-colorable. This is $O(kn)$ vertices and $O(kn)$ edges.

(b) For all $1 \leq i \leq n$ we need that *at most one* of $v_{i1}, \ldots, v_{ik}$ is colored T. Hence we need that, for each pair of vertices v_{ij_1}, v_{ij_2} at most one is colored T. For each $1 \leq j_1 < j_2 \leq k$ we place the gadget $GAD(v_{ij_1}, v_{ij_2}, F)$. This is $2n\binom{k}{2} = O(k^2 n)$ vertices and $5n\binom{k}{2} = O(k^2 n)$ edges.

(2) For each edge (v_i, v_j) in the original graph we want to make sure that v_i and v_j are not the same color. Place the gadgets $GAD(v_{i1}, v_{j1}, F)$, $GAD(v_{i2}, v_{j2}, F), \ldots,$ $GAD(v_{ik}, v_{jk}, F)$. This is $O(ke)$ vertices and $O(ke)$ edges.

Note that the number of vertices and edges in G' are both $O(k^2 n + ke)$.

Clearly G is k-colorable iff G' is 3-colorable. $\qquad \square$

We just proved $COL_k \leq COL_3$ by a simple reduction. What about $COL_a \leq COL_b$ for general a, b? If $a \leq b$ then this is easy as was explained at the beginning of this chapter. If $a > b$ then one can do either:

(1) a proof similar to that of Theorem 22.1, or
(2) use $COL_a \leq COL_3$ as in Theorem 22.1, and then do $COL_3 \leq COL_b$, which is easy.

References

Karp, R. (1972). Reducibility among combinatorial problems, in *Complexity of computer computations*, pp. 85–103.

Lovasz, L. (1973). Coverings and colorings of hypergraphs, in *Proc. of the 4th Southeastern Conference on Combinatorics, Graph Theory, and Computing*, pp. 3–12, www.cs.elte.hu/~lovasz/scans/covercolor.pdf.

Stockmeyer, L. (1973). Planar 3-colorability is polynomial complete, *SIGACT News* **5**, 1.

Chapter 23

Rectangle-Free Coloring of Grids

Stephen Fenner, Charles Glover, Christian Posthoff
Semmy Purewal, Bernd Steinbach

Prior Knowledge Needed: None.

23.1 Point

In this chapter, I take you through a line of research that I came up with and see where it leads. While there is some nice math in this chapter, the main point is to just tell a story about how math comes about and how offering big bucks can lead to interest and activity in a problem.

This chapter is based on a long paper I wrote with Stephen Fenner, Charles Glover, and Semmy Purewal [Fenner *et al.* (2012)]. For more information and proofs, see that paper. In addition, the main open problem that question raised was solved by Christian Posthoff and Bernd Steinbach [Steinbach and Posthoff (2012b,c,a)]. For more information and proofs, see that paper.

23.2 Motivation

The following theorem from Ramsey Theory is well known:

For all 2-colorings of the $\mathbb{N} \times \mathbb{N}$ *grid there is a monochromatic square (a square with all four corners the same color.).*

We can use more colors and finitize it:
For all c, there exists $W = W(c)$ *such that, for all c-colorings of* $\{1, \ldots, W\} \times \{1, \ldots, W\}$ *there exists a monochromatic square.*

The classical proof of the theorem gives very large upper bounds on $W(c)$. Axenovich and Manske [Axenovich and Manske (2008)] reduced the bound on $W(2)$. Bacher and Eliahou [Bacher and Eliahou (2011)] and (independently) Walton and Li [Walton and Li (2013)] have used computer programs to show $W(2) = 15$. Using a program for $W(3)$ seems hopeless for now. So what to do? Change the problem! We relax the problem to seeking a *monochromatic rectangle* then we can obtain far smaller bounds. In some cases exact bounds. (Our problem is equivalent to finding bipartite Ramsey Numbers but we prefer our motivation.)

This is how research often goes. Getting a better handle on when you get a monochromatic square seems hard. So we look at monochromatic rectangles. Will this help the problem of monochromatic squares? Probably not.

People often ask me *why are you working on X* (most recently, the muffin problem)? The answer may have a chicken-and-egg feel to it: I work on X. I make progress on X. The progress is interesting. Hence X is interesting. Hence I work on X...

Definition 23.1. A *rectangle* of $G(n, m)$ is a subset of the form $\{(a, b), (a + c_1, b), (a + c_1, b + c_2), (a, b + c_2)\}$ for some $a, b, c_1, c_2 \in \mathbb{N}$. A grid $G(n, m)$ is *c-colorable* if there is a function $\chi : G(n, m) \to [c]$ such that there are no rectangles with all four corners the same color.

Not all grids have *c*-colorings.

Exercise: Show that, for all $c \geq 1$, $G(c+1, c^{c+1}+1)$ does not have a c-coloring.

Definition 23.2. Let $n, m, n', m' \in \mathbb{N}$. $G(m, n)$ *contains* $G(n', m')$ if $n' \leq n$ and $m' \leq m$. $G(m, n)$ *is contained in* $G(n', m')$ if $n \leq n'$ and $m \leq m'$. Proper containment means that at least one of the inequalities is strict.

Clearly, if $G(n, m)$ is c-colorable, then all grids that it contains are c-colorable. Likewise, if $G(n, m)$ is not c-colorable then all grids that contain it are not c-colorable.

Definition 23.3. Fix $c \in \mathbb{N}$. An *Obstruction Set*, OBS_c, is the set of all grids $G(n, m)$ such that $G(n, m)$ is not c-colorable but all grids properly contained in $G(m, n)$ are c-colorable.

The following theorem is almost a tautology. Hence we do not prove it.

Theorem 23.1. *Fix $c \in \mathbb{N}$. A grid $G(n, m)$ is c-colorable iff it does not contain any element of OBS_c.*

By Theorem 23.1 we can rephrase the question
Which grids are c-colorable?
with
What is OBS_c?
This is very nice mathematically. Rather than say *what are the c-colorable grids*, we have a very nice compact way to describe them. There is also a notion of OBS_c for the original problem of monochromatic squares; however, the notion of finding even OBS_2 is thought to be hopeless. How do I know? I have never seen it mentioned.

We now offer the readers a choice. Some of you will want to learn the mathematics involved. If so, then read on. Other readers will want to skip the mathematics and get to the exciting sequence of events that involved offering money to get problems solved. For those people, skip to Section 23.7.1. I'll even have a paragraph welcoming you!

23.3   Our Main Tool to Show Grids are Not c-Colorable

We give an example of how to show a grid is not c-colorable.

Theorem 23.2. *The 21×23 grid is not 3-colorable.*

Proof. Let χ be a 3-coloring of 22×23. We show that χ must have a monochromatic rectangle. Assume the colors are R, B, G.
   KEY:
$$\text{some color was used } \left\lceil \tfrac{22 \times 23}{3} \right\rceil = 169 \text{ times.}$$
Assume the color is R.
   Since there are 23 columns there is some column with $\geq \left\lceil \tfrac{160}{23} \right\rceil = 8$ Rs. Just looking at those rows we have the following picture:

	1	2	3	4	5	6	7	8	9	10	11	12	13	14	15	16	17	18	19	20	21	22	23
1	R	?	?	?	?	?	?	?	?	?	?	?	?	?	?	?	?	?	?	?	?	?	?	?
2	R	?	?	?	?	?	?	?	?	?	?	?	?	?	?	?	?	?	?	?	?	?	?	?
3	R	?	?	?	?	?	?	?	?	?	?	?	?	?	?	?	?	?	?	?	?	?	?	?
4	R	?	?	?	?	?	?	?	?	?	?	?	?	?	?	?	?	?	?	?	?	?	?	?
5	R	?	?	?	?	?	?	?	?	?	?	?	?	?	?	?	?	?	?	?	?	?	?	?
6	R	?	?	?	?	?	?	?	?	?	?	?	?	?	?	?	?	?	?	?	?	?	?	?
7	R	?	?	?	?	?	?	?	?	?	?	?	?	?	?	?	?	?	?	?	?	?	?	?
8	R	?	?	?	?	?	?	?	?	?	?	?	?	?	?	?	?	?	?	?	?	?	?	?

Case 1: Some row has two Rs in it. Then we have a monochromatic rectangle.
Case 2: Every row has at most one R in it. Hence every row has either (a) at least four Bs, or (b) at least four Gs. Map every column in which B or G appears four times (break ties arbitrarily). We now have that there are ≥ 12 columns with

(say) four Bs. We can assume that columns 2 through 13 in the following picture have 4 Bs in them:

	1	2	3	4	5	6	7	8	9	10	11	12	13
1	R	B	?	?	?	?	?	?	?	?	?	?	?
2	R	B	?	?	?	?	?	?	?	?	?	?	?
3	R	B	?	?	?	?	?	?	?	?	?	?	?
4	R	B	?	?	?	?	?	?	?	?	?	?	?
5	R	?	?	?	?	?	?	?	?	?	?	?	?
6	R	?	?	?	?	?	?	?	?	?	?	?	?
7	R	?	?	?	?	?	?	?	?	?	?	?	?
8	R	?	?	?	?	?	?	?	?	?	?	?	?

We leave it to the reader to show that any placing of four Bs in columns 3,4 (that's as far as you need to go!) yields a monochromatic rectangle. □

The above proof used the following concept and theorem.

Definition 23.4. A *rectangle-free subset* $A \subseteq G(n, m)$ is a subset that does not contain a rectangle.

Theorem 23.3. *If $G(n, m)$ is c-colorable, then it contains a rectangle-free subset of size $\lceil \frac{nm}{c} \rceil$.*

So far *all* of our proofs that grids are not c-colorable used Theorem 23.3. Actually they use Theorem 23.3 and (as we did above) a proof that a set above a certain size cannot be rectangle-free. So Theorem 23.3 is not a magic wand — you still need to do some work to use it. We did obtain some nice theorems in the full paper that *are* a magic wand and solved a problem easily. In our exposition below we will refer to theorems that fell out of our tools and those we had to work at.

Historically this was murky. We would obtain *ad hoc* results but then develop tools from which they fell out and get more results. We still have some *ad hoc* results that we hope will lead to new tools.

Conjecture 23.1. *If $G(n, m)$ has a rectangle-free set of size $\lceil \frac{mn}{c} \rceil$ then $G(n, m)$ is c-colorable.*

23.4   Our Main Tool to Show Grids are c-Colorable

We have many constructions for colorings using finite fields and tournaments. However, they are a bit too complicated for this chapter so the interested reader can read the paper itself. For now the point is that we have such tools. In our exposition we will comment on whether they sufficed.

23.5   Which Grids are 2-Colorable?

We first considered 2-colorings. The following theorem can be proved from our tools; however, its easier than that.

Exercise: Prove the following theorem. Hint: You can do this! Real Hint: To prove that a grid is not 2-colorable use Theorem 23.3.

Theorem 23.4.

(1) The following are not 2-colorable: $G(3,7)$, $G(5,5)$, $G(7,3)$.
(2) The following are 2-colorable: $G(2,7)$, $G(4,6)$, $G(6,4)$, $G(7,2)$.

From Theorem 23.4 we can show:

Theorem 23.5. OBS_2 *consists of the following grids.*

$$\{G(3,7), G(5,5), G(7,3)\}.$$

23.6 Which Grids Can be 3-Colored?

The following theorem came from our tools. They may be difficult for the reader to do.

When we first obtained these theorems we didn't have the tools. Mathematics can be a constant back-and-forth between particular examples and general theorems.

Theorem 23.6.

(1) The following grids are not 3-colorable: $G(4,19)$, $G(5,16)$, $G(7,13)$, $G(10,12)$, $G(11,11)$, and their reversals.

(2) The following grids are 3-colorable: $G(3,19)$, $G(4,18)$, $G(6,15)$, $G(9,12)$, and their reversals.

The rest of the theorems in this chapter had to be done by special cases.

Theorem 23.7. $G(10,10)$ *is 3-colorable.*

Proof. Here is the 3-coloring:

Table 23.1 3-Coloring of $G(10,10)$.

1	1	1	1	2	2	3	3	2	3
1	2	2	3	1	1	1	3	3	2
3	1	2	3	1	2	2	1	1	3
3	2	1	2	2	1	3	1	3	1
1	2	3	3	3	2	3	2	1	1
3	1	2	2	3	3	1	2	2	1
2	3	1	2	3	2	1	3	1	2
2	2	3	1	1	3	2	3	2	1
3	3	3	1	2	1	2	2	1	2
2	3	2	1	2	3	1	1	3	3

□

Note 23.1. We found the coloring in Theorem 23.7 by the following steps.

- We found a size 34 rectangle-free subset of $G(10, 10)$ (by hand). Frankly we were trying to prove there was no such rectangle-free set and hence $G(10, 10)$ would not be 3-colorable.
- We used the rectangle-free set for color 1 and completed the coloring with a simple computer program.

It is an open problem to find a general theorem that has a corollary that $G(10, 10)$ is 3-colorable.

There is only one case left: $G(11, 10)$. We obtained that it is not 3-colorable:

Theorem 23.8. *If $A \subseteq G(11, 10)$ and A is rectangle-free then $|A| \leq 36 = \lceil \frac{11 \cdot 10}{3} \rceil - 1$. Hence $G(11, 10)$ is not 3-colorable.*

The proof of Theorem 23.8 is awful. It's long, there are lots of cases, and it does not lead to a general theorem that will help in later investigations. To be fair, *while I was working on it*, it was interesting. It required some cleverness (and hence some co-authors). To be honest, I didn't like re-reading it while working on this chapter.

The upshot is that our general theorems got us most of the way there, and we did special case proofs for $G(10, 10)$ and $G(10, 11)$, all leading to:

Theorem 23.9. OBS$_3$ *consists of the following grids and their reversals.*

$$\{G(4, 19), G(5, 16), G(7, 13), G(10, 11).\}$$

The special cases in this chapter make us wonder — will things get uglier as c gets bigger? Or will we actually develop tools to make things easier?

23.7 Which Grids Can be 4-Colored?

Using our tools we obtained the following:

Theorem 23.10.

(1) The following grids are not 4-colorable: $G(5,41)$, $G(6,31)$, $G(7,29)$, $G(9,25)$, $G(10,23)$, $G(11,22)$, $G(13,21)$, $G(17,20)$, $G(18,19)$, and their reversals.

(2) The following grids are 4-colorable: $G(4,41)$, $G(5,40)$, $G(6,30)$, $G(28,8)$, $G(9,24)$, $G(16,20)$, and their reversals.

There are just a few grids that are not covered. Here is one that we had to do separately. The proof is casework and is omitted.

Theorem 23.11. $G(19,17)$ *is not 4-colorable.*

23.7.1 *Results That Needed a Computer Program*

For the readers that just rejoined us, here is where we are: In Section 23.5, we determined OBS_2 and hence exactly which grids are 2-colorable. In Section 23.6, we determined OBS_3 and hence exactly which grids are 3-colorable. But then with 4-coloring we hit a snag.

For all our readers: At this point in the paper the only grids whose 4-colorability are unknown are $G(22,10)$, $G(21,11)$, $G(21,12)$, $G(17,17)$, $G(17,18)$, and $G(18,18)$. What to do?

This may seem like a computational problem that one could solve with a computer; however, the number of possible 4-coloring of (say) $G(18,18)$ is on the order of 4^{324}. By contrast, the number of protons in the universe, also called Eddington's number, has been estimated at approximately 4^{128}.

The only technique we know of to show that $G(n,m)$ is *not* c-colorable is to show that no rectangle-free set of $G(n,m)$ is of size $\geq \lceil \frac{nm}{c} \rceil$. In 2008 we obtained a rectangle-free set of

$G(17, 17)$ of size $74 = \left\lceil \frac{17 \times 17}{4} \right\rceil + 1$ (using a computer program). Hence either:

(1) $G(17, 17)$ is 4-colorable, or
(2) $G(17, 17)$ is not 4-colorable; however the proof of this uses a new technique since it can't use the lack of large rectangle-free sets.

While (2) would be more interesting than (1), I thought that (1) was the answer. And I wanted to know! So *how to get a math problem solved?*

One method: offer money for it. On November 30, 2009 I posted on my blog [Gasarch (2009)]

The 17×17 challenge:

If someone emails William a 4-coloring of $G(17, 17)$ then he will give them $289.00.

I also posted the (then) current version of the paper.

(1) Brian Hayes, a popular science writer, put the problem on his blog [Hayes (2009)] thus exposing the problem to many more people. (The phrase *popular science writer* is ambiguous. Does he write about popular science, that is, science that the layperson can understand? Or is he a science writer who gets invited to a lot of parties?)
(2) Brad Larsen noticed that we didn't have 4-colorings of $G(21, 11)$ and $G(22, 10)$. He then found such 4-colorings using a SAT solver which, in his words, took about 45 seconds.
(3) Many people worked on finding a 4-coloring of $G(17, 17)$ (for the money! For the glory!) but could not solve it. This led to speculation that the problem may be difficult. Apon *et al.* [Apon *et al.* (2012)] later found evidence that the problem was hard, though by that time a 4-coloring of

$G(17, 17)$ had already been found. Irony? The evidence was an NP-completeness result. This indicates that there are likely no general tools; however, that does not mean that $G(17, 17)$ in particular is hard.

(4) Bernd Steinbach and Christian Posthoff worked on solving the problem with SAT solvers. In a sequence of three brilliant papers they solved the problem [Steinbach and Posthoff (2012b,c,a)]. This was very serious and deep research that may lead to improved SAT Solvers for other problems. They announced their result in February of 2012 and I blogged about it [Gasarch (2012)]. Dr. Gasarch happily paid them the $289.00.

(5) Marzio De Biasi easily found an extension of the 4-coloring of $G(17, 17)$ to $G(18, 18)$ and posted it as a comment on that blog. Bernd Steinbach and Christian Posthoff had already known this coloring as well. Neither one demanded $324.00.

(6) Bernd Steinbach and Christian Posthoff used their techniques to find a 4-coloring of $G(21, 12)$ and posted it as a comment on the blog post [Gasarch (2012)]. With this OBS_4 was completely known!

(7) Inspired by the 17×17 challenge and the solution to it Neil Brewer and Dmitry Kamenetsky devised a contest at http:infinitesearchspace.dyndns.org that asked for the following: *For $c = 1$ to 21 find the largest n such that the $n \times n$ grid is c-colorable. You must also present the coloring.* This lead to a lot of interesting discussion including the following two points, one of which we used in our paper

(a) Tom Sirgedas obtained another 4-coloring of $G(21, 12)$. To paraphrase him: *I noticed that the rectangle-free subset A of $G(21, 12)$ in (the earlier version of) the paper had the following property: If you viewed it as a 7×3 grid of 3×3 grids then in each of those 3×3 grids either all elements of the diagonal were all in A or none* *were in A. I assumed that the solution would have this property. This cut down the number of possibilities by quite a lot. Then, I just wrote an exhaustive depth-first-search to fill the grid one color at a time, and each color one row at a time. I used a lot of pruning and bitmasks, and solutions were found in a few minutes. Unfortunately this approach seems to only work for this particular grid. It won't scale well at all.*

(b) Quimey Vivas posted a proof that if c is prime then $G(c^2, c^2)$ is c-colorable. Ken Berg had previously sent me a proof that $G(c^2, c^2 + c)$ is c-colorable when c is a power of a prime.

Theorem 23.12. $G(21, 12)$ *is 4-colorable*

Proof. Bernd Steinbach and Christian Posthoff (as a team) and Tom Sirgedas obtained a 4-coloring of $G(21, 12)$:

□

Theorem 23.13. $G(22, 10)$ *is 4-colorable*

Proof. Brad Larsen obtained a 4-coloring of $G(22, 10)$:

□

Theorem 23.14. $G(18, 18)$ *is 4-colorable*

Proof. Bernd Steinbach and Christian Posthoff obtained a 4-coloring of $G(18, 18)$.

□

Finally from this we have:

Theorem 23.15. OBS_4 *consists of the following grids and their reversals:*

$$\{G(5, 41), G(6, 31), G(7, 29), G(9, 25),$$

$$G(10, 23), G(11, 22), G(13, 21), G(17, 19)\}$$

Table 23.2   A 4-coloring of $G(21,12)$ due to Tom Sirgedas.

1	2	2	3	2	1	3	4	4	3	1	3
2	1	2	1	3	2	4	3	4	3	3	1
2	2	1	2	1	3	4	4	3	1	3	3
3	4	2	4	1	1	1	2	4	3	2	4
2	3	4	1	4	1	4	1	2	4	3	2
4	2	3	1	1	4	2	4	1	2	4	3
2	4	1	3	2	4	2	3	3	4	2	1
1	2	4	4	3	2	3	2	3	1	4	2
4	1	2	2	4	3	3	3	2	2	1	4
3	1	1	2	2	1	3	4	1	4	3	4
1	3	1	1	2	2	1	3	4	4	4	3
1	1	3	2	1	2	4	1	3	3	4	4
3	4	4	1	3	2	2	4	2	1	1	3
4	3	4	2	1	3	2	2	4	3	1	1
4	4	3	3	2	1	4	2	2	1	3	1
3	4	2	3	4	3	2	1	1	1	4	2
2	3	4	3	3	4	1	2	1	2	1	4
4	2	3	4	3	3	1	1	2	4	2	1
3	3	1	4	2	4	4	1	3	2	1	2
1	3	3	4	4	2	3	4	1	2	2	1
3	1	3	2	4	4	1	3	4	1	2	2

23.8   It's a Floor Wax! It's a Dessert Topping! It's Both!

So is the study of grid colorings math or computer science? The theorems about colorings are clearly math. The improvements to SAT Solvers used to solve some of our open problems are clearly computer science. While I am happy to have specific 4-colorings I wish we had more general theorems from which the 4-colorings we needed would fall out. Why? Because I doubt computing power will be enough for 5-colorings.

Is it a good math problem? Only time will tell, but for now progress seems to be in the realm of computer science.

Table 23.3 A 4-coloring of $2_{22,10}$ due to Brad Larsen.

1	2	3	3	2	2	1	1	4	4
2	1	4	2	4	4	1	3	1	3
4	2	4	3	1	1	2	3	1	4
1	1	2	2	3	3	4	4	2	1
1	4	1	1	2	3	3	2	2	3
1	4	3	4	3	1	2	3	2	2
2	1	2	1	4	1	3	4	3	2
1	3	3	4	1	4	2	2	4	3
1	4	4	3	3	2	3	2	1	2
3	3	4	4	1	2	3	4	2	1
3	2	2	1	3	4	4	2	1	3
3	4	3	2	2	1	1	4	4	2
4	3	2	4	2	3	4	3	1	1
2	2	1	4	4	1	3	3	2	4
3	2	1	3	4	3	4	1	1	2
4	4	1	2	1	4	1	2	3	3
2	1	4	3	1	2	4	1	4	3
3	4	2	1	4	2	1	3	3	1
2	4	3	1	1	3	4	2	3	4
4	3	1	2	3	2	2	4	3	1
4	3	2	3	4	1	2	1	4	1
2	1	1	3	2	4	2	4	3	4

23.9 Open Questions

(1) Refine our tools so that our ugly proofs can be corollaries of our tools.

(2) Find an algorithm that, given c, will quickly find OBS_c or $|\mathrm{OBS}_c|$.

(3) We know that $2\sqrt{c}(1 - o(1)) \le |\mathrm{OBS}_c| \le 2c^2$. Bring these bounds closer together.

(4) All of our results of the form $G(n, m)$ *is not c-colorable* have the same type of proof: show that there is no rectangle-free subset of $G(n, m)$ of size $\lceil ab/c \rceil$. Either

- show that if a grid $G(n, m)$ has a rectangle-free set of size $\lceil nm/c \rceil$ then it is c-colorable, or

Table 23.4 A 4-coloring of $2_{18,18}$ due to Bernd Steinbach and Christian Posthoff.

1	2	2	1	4	4	4	1	3	1	1	3	4	3	2	3	4	2
3	1	4	4	4	3	3	2	4	1	2	2	1	1	2	3	2	3
2	2	3	3	1	1	4	2	4	2	1	4	1	3	4	4	1	3
1	1	3	2	4	3	1	2	1	4	4	3	2	4	3	4	1	2
2	4	2	3	3	4	3	3	4	1	2	3	2	4	1	2	1	1
3	4	4	1	1	2	2	2	1	4	3	3	3	1	4	2	4	1
2	1	3	2	2	2	3	4	3	3	1	4	3	4	2	1	4	1
4	1	4	3	1	2	4	1	2	2	2	1	3	4	3	3	3	2
4	4	1	3	4	3	2	1	2	3	3	4	2	1	2	1	1	4
2	3	3	4	3	4	2	1	1	4	3	4	1	2	1	3	2	2
4	1	1	1	2	1	3	4	4	4	3	2	4	3	1	2	3	2
3	2	3	4	2	1	2	3	1	1	2	1	4	4	4	1	3	4
3	2	4	2	3	1	1	1	2	3	4	4	4	3	3	2	2	1
3	3	4	3	2	4	1	4	3	2	1	1	2	1	1	4	2	4
4	3	2	1	2	4	1	2	2	3	4	3	1	2	4	1	3	3
1	3	2	2	1	3	2	3	4	2	4	2	3	3	1	1	4	4
1	4	1	4	3	3	4	4	3	2	4	1	1	2	2	2	3	1
4	2	1	4	1	2	1	3	3	1	3	2	2	2	3	4	4	3

- develop some other technique to show grids are not c-colorable.

(5) Find OBS_5 and beyond!

References

Apon, D., Gasarch, W., and Lawler, K. (2012). An NP-complete problem in grid coloring, http://arxiv.org/abs/1205.3813.

Axenovich, M. and Manske, J. (2008). On monochromatic subsets of a rectangular grid, *Integers* **8**, 1, p. A21, http://orion.math.iastate.edu/axenovic/Papers/Jacob-grid.pdf and http://www.integers-ejcnt.org/vol8.html.

Bacher, R. and Eliahou, S. (2011). Extremal binary matrices without constant 2-squares, *Journal of Combinatorics* **1**, 1, pp. 77–100, `http://dx.doi.org/10.4310/JOC.2010.v1.n1.a6`.

Fenner, S., Gasarch, W., Glover, C., and Purewal, S. (2012). Rectangle-free colorings of grids, `http://arxiv.org/abs/1005.3750`.

Gasarch, W. (2009). The 17×17 challenge. Worth \$289.00. This is not a joke, `http://blog.computationalcomplexity.org/2009/11/17x17-challenge-worth-28900-this-is-not.html`.

Gasarch, W. (2012). The 17×17 SOLVED! also 18 × 18 `http://blog.computationalcomplexity.org/2012/02/17x17-problem-solved-also-18x18.html`.

Hayes, B. (2009). The 17×17 challenge, `http://bit-player.org/2009/the-17x17-challenge`.

Steinbach, B. and Posthoff, C. (2012a). Extremely complex 4-colored rectangle-free grids: Solution of an open multiple-valued problem, in *Proceedings of the Forty-Second IEEE International Symposia on Multiple-Valued Logic*, `http://www.cs.umd.edu/~gasarch/PAPERSR/17solved.pdf`.

Steinbach, B. and Posthoff, C. (2012b). The solution of ultra large grid problems, in *21st International Workshop on Post-Binary USLI Systems*, `http://www.informatik.tu-freiberg.de/index.php?option=com_content&task=view&id=35&Itemid=63`.

Steinbach, B. and Posthoff, C. (2012c). Utilization of permutation classes for solving extremely complex 4-colorable rectangle-free grids, in *Proceedings of the IEEE 2012 international conference on systems and informatics*, `http://www.informatik.tu-freiberg.de/index.php?option=com_content&task=view&id=35&Itemid=63`.

Walton, P. and Li, W. N. (2013). Searching for monochromatic-square-free Ramsey grid colorings via SAT solvers, in *International conference on information science and applications (ICISA 2013)*, p. A21.

Chapter 24

If That Were True I Would Know It! A Result in Kolmogorov Complexity

Prior Knowledge Needed: Helpful but not required to know (1) the halting problem, (2) Turing reductions, and (3) the definition of Kolmogorov complexity.

24.1 Point

As a graduate student in 1983, I came across a book in the library that claimed to solve Fermat's Last Theorem (FLT) using elementary methods. (This was before Wiles showed FLT true using rather advanced methods.) I **knew** that the claim was false. Why? Because **if FLT had really been solved, I would know it**. I was not bragging about how much math I knew; I was bragging about how much math I knew *of*. Was that a rigorous proof technique? No, but it is useful and often correct.

I now tell the tale of how I used this technique in Kolmogorov Complexity. In particular, I **knew** that the Kolmogorov function could not be of intermediary Turing degree because **if there were a natural intermediary Turing degree then I would know that**. This chapter will be self-contained; hence, I will define all of these terms. For more on Kolmogorov complexity see the classic book by Li and Vitányi [Li and Vitányi (2008)].

24.2   How to Measure Randomness Intuitively

Intuitively, the string 00000000000000000000000 does not seem random. How to make this rigorous? Note that there is a program of length $\lg n + O(1)$ that prints out 0^n:

for $i = 1$ to n print(0)

Conversely, the string 011010001100000011101010101000110 0 does seem random. The shortest program to print it out might very well be

print(0110100011000000111010101000110 0)

which is roughly the length of the string itself.

Taking a cue from the above two examples, we will define the *randomness of a string* x to be the size of the shortest program that prints x. We need a formal notion of program.

24.3   A Short Introduction to Computability Theory

We will not define Turing machines formally. All you need to know is the following:

- A function can be computed by a Turing machine iff the function can be computed by a computer program, say, written in Java.
- Much like Java programs, it is quite possible that a Turing machine run on an input will never halt.
- Turing machines have a finite description, which can be interpreted as code. That is, given a description M and an input x, one can run $M(x)$.
- The program
  For $i = 1$ to n print(0)
  has length $\lg n + O(1)$.

- If x is a string of length n then the program
$$\texttt{print}(x)$$
  has length $n + O(1)$.

Notation 24.1. Let $M_1, M_2, \ldots$ be a list of all Turing machines. Let $i, x \in \mathbb{N}$. If, when machine M_i is run on input x, the computation terminates with output y, then we write $M_i(x) = y$. We take the index i to be in base 2 so the length of M_i is $|i| = \lg(i) + O(1)$.

Definition 24.1. A function $f : \mathbb{N} \to \mathbb{N}$ is *computable* if there exists a Turing machine M such that, for all x, $M(x) = f(x)$. A set A is *decidable* if its characteristic function is computable.

Example 24.1. Most functions you encounter in mathematics are computable. For example, addition, squaring, exponentiation, and given n find the nth prime. Most sets you encounter in mathematics are decidable. For example, evens, primes, and squares.

Definition 24.2. Let $i, s, x \in \mathbb{N}$.

$$M_{i,s}(x) = \begin{cases} y & \text{if the computation } M_i(x) \text{ halts within} \\ & s \text{ steps and outputs } y \\ \text{NO} & \text{otherwise} \end{cases}$$

(24.1)

The function $f(i, s, x) = M_{i,s}(x)$ is computable: Run $M_i(x)$ for s steps and either (1) it halts within s steps, so output what it outputs, or (2) it does not, so output NO.

We now give an example of an undecidable set.

Definition 24.3. HALT is the set $\{x : M_x(x) \text{ halts}\}$.

Note 24.1. A more natural set for HALT would be $\{(x, y) : M_x(y) \text{ halts}\}$. However, the version we use is easier to work with and ends up being equivalent, as we will see later.

Theorem 24.1. HALT *is undecidable.*

Proof. Assume, by way of contradiction, that HALT is decidable. Hence we will be able to make the query $x \in$ HALT in our program and get an answer back. Let M_i be the Turing machine that does the following

(1) Input(x)
(2) Ask if $x \in$ HALT. Note that this will be YES if $M_x(x)$ halts and NO if $M_x(x)$ does not halt.
(3) If $x \in$ HALT then go into an infinite loop, else halt.

Is $i \in$ HALT? We show that both answers, YES and NO, lead to a contradiction.

If $i \in$ HALT then $M_i(i)$ halts. Hence $i \in$ HALT. When M_i is run on i, the instructions in step 3 say that it goes into an infinite loop. Hence $i \notin$ HALT. This is a contradiction.

If $i \notin$ HALT then $M_i(i)$ does not halt. Hence $i \notin$ HALT. When M_i is run on i, the instructions in step 3 say that the computation halts. Hence $i \in$ HALT. This is a contradiction.

$\square$

There are many ways to express the above proof. My favorite is Geoffrey Pullum's poem in the style of Dr. Seuss [Pullum (2004)].

We need the notion of *If I had access to the function g then the function f would be computable.* We could define oracle Turing machines formally but instead we define it informally.

Definition 24.4. An *Oracle Turing Machine* is a Turing machine that has the ability to make a call to an unspecified function g, called *the oracle*. The Oracle Turing machine is defined independently of the function. We denote an oracle Turing machine by $M_i^{()}$. The notation $M_i^g(x)$ means that we run the $M_i^{()}$ with oracle g and on input x.

Definition 24.5. Let $f, g : \mathbb{N} \to \mathbb{N}$. We say $f \leq_T g$ if there is an oracle Turing machine $M^{()}$ such that M^g computes f. This is called a *Turing reduction*. We say $f <_T g$ if $f \leq_T g$ but $g \not\leq_T f$, and $f \equiv_T g$ if $f \leq_T g$ and $g \leq_T f$.

Exercise Let $\text{HALT}_0 = \{(x, y) \mid M_x(y) \text{ halts}\}$. Show that $\text{HALT} \leq_T \text{HALT}_0$ and $\text{HALT}_0 \leq_T \text{HALT}$. This is what we meant in the note following Definition 24.1 by saying that the two sets were equivalent.

I leave the following lemma to the reader:

Lemma 24.1. *If $f \leq_T g$ and g is computable then f is computable.*

Given a function, how do you prove that is undecidable? One way is by contradiction. The proof that HALT is undecidable is by contradiction. We will later define the Kolmogorov function and show it is not computable by contradiction.

The most common way to show that a function f is not computable is to take a known undecidable set A (almost always HALT) and show $A \leq_T f$.

24.4 Intermediary Sets

We can rewrite the statement HALT *is undecidable* as $\emptyset <_T$ HALT. A set A such that $\emptyset <_T A <_T$ HALT is called an *intermediary sets*. In 1950, it was an open problem to determine if there were any such sets. In 1954, Post and Kleene first showed there were such sets. What are these sets? I once had the following conversation with my Darling:

Bill: Darling, if I told you there were sets that were not decidable but easier than the halting problem, would you find that interesting?

Darling: Yes, unless...

Bill: Great! Because there are. Oh, I interrupted you
              — what were you saying?

Darling: I would find the existence of such sets interesting
              unless they were some dumb-ass sets that people
              constructed *just* for the sole point of being unde-
              cidable but easier than the halting problem.

Bill: You nailed it!

In a nutshell, such sets are not natural. Hence the sets themselves are not interesting. It would be nice to have natural, and
hence interesting, intermediary sets.

Most computability theorists think that there are no natural
intermediary sets. It is not clear how to prove this. In fact, it is
not even clear how to state this rigorously.

24.5   How to Measure Randomness Formally

We use TM to abbreviate "Turing machine."

Definition 24.6. The *Kolmogorov complexity of a string* x, denoted $C(x)$, is the length of the shortest TM M such that
$M(0) = x$. We often call a TM that prints out a string x a
description of x.

Note 24.2. The definition of C depends on the formal definition
of TM. It is possible that if you define TMs using Java programs
then $C(0^n) = \lfloor \lg(n) \rfloor + 1000$ whereas if you define TMs using
Fortran programs then $C(0^n) = \lfloor \lg(n) \rfloor + 5$. However, for any
two types of TMs, there is a constant c such that the two $C(x)$s
differ by at most c. Hence this will not affect any of our results.

C is a measure of randomness. If $C(x) \geq |x|$ then we think
of x as being random. Are there such strings? Yes!

Lemma 24.2. *For all $n \geq 1$ there is a string $x \in \{0,1\}^n$ such that $C(x) \geq n$.*

Proof. Assume, by way of contradiction, for all $x \in \{0,1\}^n$, $C(x) < n$. Map each $x \in \{0,1\}^n$ to the program that prints it. Note that this map is 1-1. There are 2^n elements in the domain and $\sum_{i=0}^{n-1} 2^i = 2^n - 1$ in the range. Hence the map cannot be 1-1. Contradiction. $\qquad\square$

How hard is C? We first show that $C \leq_\mathrm{T}$ HALT and then that C is not computable.

Theorem 24.2. $C \leq_\mathrm{T}$ HALT.

Proof. Let c be the constant such that, for all x, $C(x) \leq |x| + c$.

(1) Input x. We want to know $C(x)$.
(2) For all programs M of length $\leq |x| + c$, ask the HALT oracle: *Does $M(0)$ halt and output x?*
(3) Output the length of the shortest M such that $M(0)$ halts and outputs x.

$\qquad\square$

Theorem 24.3. C *is not computable.*

Proof. Assume, by way of contradiction, that C is computable. Assume also that the program for C is of size s. Consider the following program (where a is a constant whose value will be determined later).

```
for each x ∈ Σᵃˢ
    compute C(x)
    if C(x) ≥ as then print(x) and stop.
```

This program is of size $s + \lg(as) + O(1)$. Its output is a string of length as. Choose a large enough so that

$$s + \lg(as) + O(1) < as.$$

But now the output is a string x such that $C(x) \geq as$ and yet it has a description of length less than as. Contradiction. $\square$

24.6 Is the Kolmogorov Function Intermediary?

As a grad student, in 1984, I came across Theorems 24.2 and 24.3 which together imply $\emptyset <_{\mathrm{T}} C \leq_{\mathrm{T}}$ HALT. Unlike most proofs of non-computability, the proof that C was not computable *did not* show $HALT \leq_{\mathrm{T}} C$. Hence it was possible that C is an intermediary set. I **knew** that possibility was false. Why? Because **if there were a natural intermediary set, I would know it.** Similar to earlier when I was not bragging about how much *mathematics* I knew; I was bragging about how much mathematics I knew of, now I am not bragging about how much *computability theory* I knew; I am bragging about how much computability theory I knew *of*. Is that a rigorous proof technique? No, but it is useful and often correct.

Was I correct? Yes. I asked around (this was before the Web, and even email was not as common in 1984 as it is in 2018) and eventually Peter Gacs (a professor at Boston University) gave a proof that HALT $\leq_{\mathrm{T}} C$ (on paper) to Mihai Gereb (a graduate student at Harvard) who gave it to me. I would have preferred to be wrong and to have seen a natural intermediary set. Oh well.

Here is the proof:

Definition 24.7. Let $C_s(x)$ be the size of the shortest TM M such that $M(0) = x$ and halts within s steps.

Theorem 24.4. HALT $\leq_{\mathrm{T}} C$.

Proof. Here is the algorithm for HALT that uses C as an oracle. The constant a will be determined later.

(1) Input(x) (we want to know if $M_x(x)$ halts). Let $|x| = n$.

(2) Find s_0 such that, for all $y \in \{0,1\}^{an}$, $C_{s_0}(y) = C(y)$. (This step uses the oracle for C.)

(3) Run $M_x(x)$ for s_0 steps. If it halts, then output YES. If not, then output NO. (We still need to prove that this is correct.)

We need to show that if $M_x(x)$ does not halt within s_0 steps then it never halts. Assume, by way of contradiction, that $M_x(x)$ halts in $s \geq s_0$ steps. Note that, for all $y \in \{0,1\}^{an}$, $C_s(y) = C(y)$. The following algorithm will be a short description of a string that lacks a short description.

(1) Run $M_x(x)$. Let s be the number of steps it took to halt.

(2) For all $y \in \{0,1\}^{an}$ compute $C_s(y)$ (which is $C(y)$).

(3) Let y be a string of length an such that $C_s(y) \geq |y|$.

(4) Output y.

The above algorithm can be described with $n + \lg(a) + O(1)$ bits. Hence

$$C(y) \leq n + \lg(a) + O(1).$$

By the definition of s we have

$$C(y) = C_s(y) \geq |y| = an.$$

Hence

$$an = |y| \leq C(y) \leq n + \lg(a) + O(1)$$

Choose a such that

$$n + \lg(a) + O(1) < an.$$

This yields a contradiction. □

24.7   The Point Reiterated

Let Q be a statement in mathematics that is either true or false. You don't know which. Should you try to prove Q or $\neg Q$? How do you decide what to spend more time on? You can do examples or see what fits in with other mathematics. You can see what people who have worked on the problem think. And you can sometimes use:

If Q is true, I would know it. Hence I will try to prove ¬Q.

I am not claiming this as a rigorous proof technique. But it may be used as a starting point. Note that I might have wasted a lot of time trying to understand an allegedly easy proof of FLT or trying to show that C was intermediary had I not employed this method.

References

Li and Vitányi (2008). *An introduction to Kolmogorov complexity and its applications* (Springer, New York, Heidelberg, Berlin), this is the third edition.

Pullum, G. (2004). Scooping the loop snooper, `http://www.lel.ed.ac.uk/~gpullum/loopsnoop.html`.

Index

$n + 1$, 3

Aaronson, Scott, x
Ackerman, Wilhelm, 26
Alliance sets, 91
Alliance-Enemy Descriptors
  (AEDs), 91
Ambrose, Alice, 27
anti-semitism, 23, 206

baseball, 13
batting average (baseball), 15
Baudet's conjecture, 62
Beigel, 214
Bernays, Paul, 27
Beth, Willem Evert, 27
binomial coefficients, 222
black hole numbers, 74
boomerang fractions, 75
Borwein integrals, 50
brain injuries and health, 72
Brouwer, L.E.J. "Notorious", 27
Byzantine General's Problem, 59

Cantor, Georg, 26, 27
Carnap, Rudolph, 28
Caroll, Lewis, 68, 73
Catalan numbers, 221

Chandra, 214
Chemical Pi, 70
Church, Alonzo, 28
computability theory, 43, 108, 258
continuum hypothesis, The
  independence of, 117
Cook-Levin Theorem, 65
Curry, Haskell, 28

Darling, 67, 155, 261
Deterministic finite automata
  (DFA), 109
dump your tables, 187

Ebert, Roger, 8
Egyptian Fraction, 144
Emerson, Ralph Waldo, vi
Enemy sets, 91
Erdös distinct distance problem,
  112
Étienne, Charles-Guillaume, 233
Euclid-Euler Theorem, 57, 227
Euclidean geometry, 120
Euler, Mathematical object named
  after, 56
Euler-Fermat Theorem, 57

Fandom, 19